Teacher's Manual

Exploring Design, Technology, & Engineering

R. Thomas Wright
Professor Emeritus, Industry and Technology
Ball State University
Muncie, Indiana

Ryan A. Brown
Assistant Professor
Illinois State University
Normal, Illinois

Publisher
The Goodheart-Willcox Company, Inc.
Tinley Park, Illinois
www.g-w.com

Manufactured in the United States of America.

Teacher's Manual ISBN 978-1-60525-423-4
Teacher's Resource CD ISBN 978-1-60525-424-1

1 2 3 4 5 6 7 8 9 – 12 – 16 15 14 13 12 11

The Goodheart-Willcox Company, Inc. Brand Disclaimer: Brand names, company names, and illustrations for products and services included in this text are provided for educational purposes only and do not represent or imply endorsement or recommendation by the author or the publisher.

The Goodheart-Willcox Company, Inc. Safety Notice: The reader is expressly advised to carefully read, understand, and apply all safety precautions and warnings described in this book or that might also be indicated in undertaking the activities and exercises described herein to minimize risk of personal injury or injury to others. Common sense and good judgment should also be exercised and applied to help avoid all potential hazards. The reader should always refer to the appropriate manufacturer's technical information, directions, and recommendations; then proceed with care to follow specific equipment operating instructions. The reader should understand these notices and cautions are not exhaustive.

The publisher makes no warranty or representation whatsoever, either expressed or implied, including but not limited to equipment, procedures, and applications described or referred to herein, their quality, performance, merchantability, or fitness for a particular purpose. The publisher assumes no responsibility for any changes, errors, or omissions in this book. The publisher specifically disclaims any liability whatsoever, including any direct, indirect, incidental, consequential, special, or exemplary damages resulting, in whole or in part, from the reader's use or reliance upon the information, instructions, procedures, warnings, cautions, applications, or other matter contained in this book. The publisher assumes no responsibility for the activities of the reader.

Contents

Introduction

As the world progresses into the new millennium, people in many countries are at a similar turning point. They see society around them plunging into the Information Age brought about by new communication and transportation systems. The world has become a global village, with each nation economically interconnected and affected by events everywhere.

Preparing young people to participate in this global, technology-driven society requires new types of educational programs. To meet this challenge, technology education is becoming an important part of the general education of all youth. Ideally, development of an understanding of technology and its impacts should start in the lower elementary grades. Actual practice, however, finds most students are being introduced to these fundamental concepts beginning in sixth or seventh grade. This instructional system—the textbook, Tech Lab Workbook, and Teacher's Manual—has been designed to present a broad, introductory, hands-on presentation of technology to young learners in this age group. The system is designed to help students develop technological literacy, which can be described as an ability to locate, interpret, and use information about tools, materials, and processes.

Technology—A Societal Universe

Technological literacy and its associated competencies were the basis for the fundamental changes that occurred during the Industrial Age and are also the basis for the changes now happening in the Information Age. Yet, few people can define and describe this pervasive phenomenon we call *technology*. Technology is the product of the human mind and its engineering spirit. This phenomenon is, therefore, the result of human volition and control. All technology has been developed by people to satisfy human needs and wants and is subject to human monitoring. Technology is not waiting to be discovered, but rather to be created and made available for people to use.

The phenomenon is the application of knowledge, tools, and skills to solve problems and extend human capabilities. This can be easily described as the work of engineers, inventors, and technologists in applying knowledge to create and commercialize devices and systems to meet human goals. Technology is human knowledge and practices that use tools, materials, and systems to produce artifacts (human-made things) and other outputs (such as pollution and scrap), in order to modify or control the natural and human-made environments. See **Figure IN-1.**

Technology	human knowledge and practices that use tools, materials, and systems to produce human-made things and other outputs in order to modify or control the natural and human-made environments.

Figure IN-1

Technology Education: A Background

Education about the practical world is as old as humankind. Early experiences were almost always trade oriented. In the last 100 years, educators have recognized that broad, general (not job-specific) education about tools and materials is important. Out of this belief, a new program took form—first as manual training, and later as industrial arts. These subjects, however, drew their content almost totally from the skilled trades and focused on learning "basic" skills and constructing take-home projects. The technology education discipline has emerged to replace the skill-oriented, technical instruction of industrial arts. This type of education helps students develop technological literacy through the study of the creation, use, and behavior of technical systems (tools, machines, materials, and techniques) and the interrelationship of these elements and systems with humans, societies, and the environment. A technology-education program with this focus helps students of all ages to be able to do the following:

- ❑ Know and appreciate the importance of technology to people and society.
- ❑ Use tools, materials, and processes effectively and safely.
- ❑ Apply designing and problem-solving processes.
- ❑ Understand the various contexts in which technology is applied.
- ❑ Deal with the forces shaping the future.

Technology and Industry

This instructional program deals with two major universals found in all societies—technology and industry. As already described, technology includes all materials, products, and systems humans have developed to extend their potential for modifying and controlling the natural and human-made environments. Technology is knowledge of efficient and appropriate action, knowledge of doing. This knowledge includes all human-made systems that allow us to grow, harvest, and process foods; communicate information and ideas; convert and apply energy; manufacture goods; construct buildings and civil structures; prevent and cure illnesses and injuries; and transport goods and people.

Industry, on the other hand, is part of a fundamental societal institution. Humans, over time, have developed five basic institutions to meet their needs. These are the family, religion, education, government, and the economy. Industry is part of the economic institution, which provides the goods and services making our lives more comfortable.

Systems

Technology and industry are systems that integrate parts into structures that behave in predictable ways. These systems have five major parts. See **Figure IN-2:**

1. Inputs (resources).
   A. People.
   B. Materials.
   C. Capital (tools).
   D. Information.
   E. Energy.
   F. Finances.
   G. Time.
2. Processes.
   A. Production.
   B. Management.

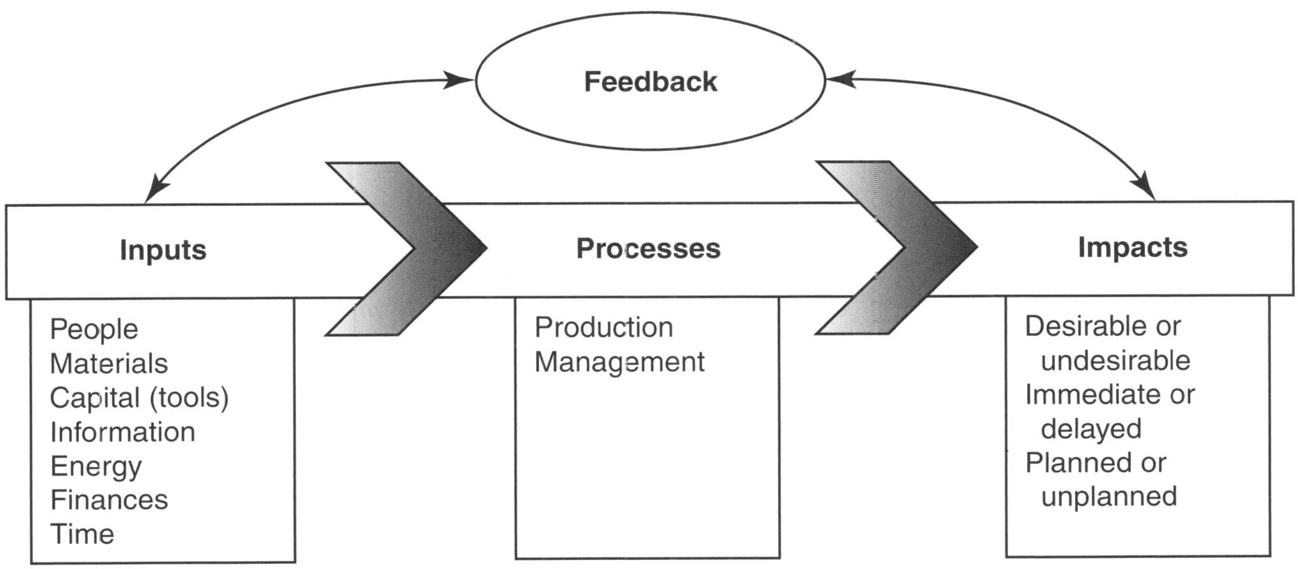

Figure IN-2

3. Outputs.
   A. Desired.
   B. Ancillary.
4. Feedback.
5. Impacts.
   A. Desirable or undesirable.
   B. Immediate or delayed.
   C. Planned or unplanned.

System Inputs

- ❑ People are fundamental to all technological systems. Only through a desire to meet human needs and wants have these systems developed. People provide the need for the system and also bring human talent to the system. They direct and operate these systems by their knowledge, skills, and attitudes.
- ❑ Materials are the resources found in nature. We use them to make products, construct structures, and power our many activities. The supplies of many resources are limited and, therefore, must be used wisely.
- ❑ Capital is the physical means of technology. This means includes all buildings, machines, tools, and equipment needed to house and operate technological systems.
- ❑ Information is the human-produced resource for all technological systems. People generate, retrieve, and apply it. Information can be cataloged under three types of knowledge. Scientific knowledge describes the natural world and provides an understanding of the natural phenomena impacting human life and activities. Technological knowledge describes the human-made world and is used to develop systems extending human potential. Humanities knowledge allows people to assess the value and need for human-made systems and interactions.
- ❑ Energy is a unique natural resource that allows work to be accomplished. Similar to materials, many energy sources are finite and must be carefully used.
- ❑ Finances are part of a system of exchange for human time, effort, and knowledge; material and energy resources; and capital goods. Money is used in modern societies to facilitate this exchange.
- ❑ Time is the measurable period during which an action, such as harvesting crops, making products, constructing buildings, using energy, or moving cargo, takes place.

System Processes

Each technological system is a dynamic interaction of practices or actions. Production processes are basic to these systems. These processes are used to change inputs into outputs. Agricultural processes transform seeds, plants, and animals into foods and other products. Communication processes transform information into media messages. Construction processes convert materials and manufactured goods into buildings and other structures. Energy-conversion processes change energy resources into heat, light, sound, and motion. Medical processes use medicines and devices to treat illnesses and promote health. Manufacturing processes change materials into goods. Transportation processes use vehicles and equipment to move people and goods. Management processes are also important to technological endeavors. They involve setting goals and courses of action (planning), structuring the resources (organizing), directing and supervising transformation activities or other work (actuating), and monitoring results and making corrective actions (controlling).

System Outputs

All human-made systems have a reason to exist. People established them to meet goals. These goals might be products, structures, media messages, services, or changed conditions. We can call these visible elements the *desired outputs*. Every system, however, also has other outputs. Technological systems generate scrap (usable materials), waste (unusable materials), and pollution (environmentally destructive materials and conditions). These ancillary outputs must be reduced or eliminated, if possible.

Feedback

Feedback is the information used to monitor and adjust technological systems so they are operating at their highest efficiencies. This information monitors the desired and ancillary outputs to determine the efficiencies of the production and management processes. Feedback indicates what changes are needed to ensure the system is adequately using system inputs.

Impacts

All technology was developed with a purpose in mind. Therefore, each product of technology was designed to have an impact. Impacts are the effects of technology on people, society, and the environment. Most technology is developed to make people's lives better. People might live in more comfort, have more information available, or move across the globe with more ease by using technologies. These are examples of positive, immediate impacts. Not all impacts, however, are positive. The use of technology can pollute the environment, displace workers, or create health risks. These are negative, undesirable impacts.

Not all impacts are immediate. Some take years to become apparent. For example, many people blame television for contributing to several societal problems, including increased violence, a declining ability to communicate verbally, and reduced creative ability in our youth. If these can rightfully be attributed to television, they are delayed, undesirable, unplanned impacts. It is easy to see that impacts of technology can be assessed from three major perspectives: desirable or undesirable, planned or unplanned, and immediate or delayed.

Organizing the Content

Many courses in schools are built on a conceptual view of a specific body of knowledge, a set of desired outcomes, and a perception of the role of the subject matter in a student's education. Teachers, curriculum writers, and instructional-material developers have individual views of their content areas. Hopefully, these views converge in a workable structure that will help students learn about the fields of study. Whatever view is used, it should meet some common criteria.

First, the content should have a logical organization that is apparent to the learner. This allows students to understand the relationship between segments of the course and helps them transfer the learning to other subjects and life situations. Second, the overall material for the course should be organized into discrete, stand-alone sections. This allows students who enter the course after it starts to easily join the class. Third, each section should present selected content and have meaningful hands-on, minds-on activities. This allows students to apply information they learn to authentic activities. Last, instruction should allow students to develop knowledge and problem-solving skills. This provides the foundation for lifelong learning.

With these principles in mind, *Exploring Design, Technology, & Engineering* is organized into five basic sections, as shown in **Figure IN-3:**

1. The first chapters introduce students to the scope of technology as a human-developed system that extends people's potential. We then discuss the system components (inputs, processes, outputs, and feedback).
2. The next section introduces the resources of technology, with chapters on tools, materials, energy, information, and people, time, and money.
3. The third section presents the design process as a way of creating technology and explores each step designers use in developing products and technological systems.
4. The fourth section explores common contexts in which technology is found. These include agricultural, construction, energy conversion, information and communication, manufacturing, medical, and transportation technologies.
5. The last section explores the role of technology in people's lives and the future of technology.

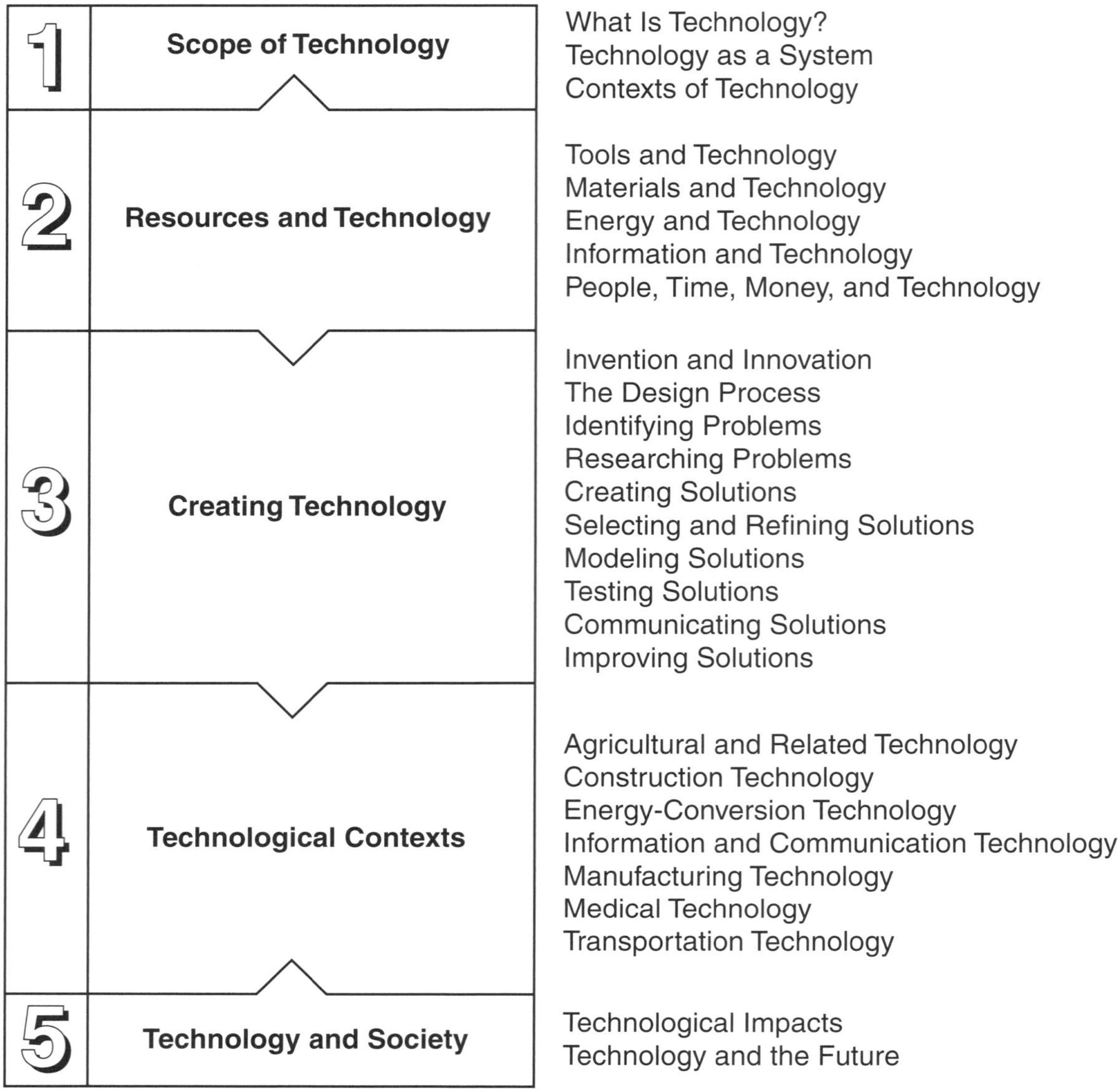

Figure IN-3

Each chapter presents a set of objectives, a well-illustrated, easily read presentation of basic information, a set of key terms, some activity ideas, and study questions. Throughout the text are fully developed activities challenging students to apply their knowledge. Additional study questions and activities are included in the Tech Lab Workbook.

Scheduling the Program

One of the most difficult tasks every teacher faces is scheduling content presentations, laboratory activities, and evaluation sessions meeting the students' needs and time constraints of the course. Curriculum writers and instructional-material developers can provide basic guidelines. By their very nature, however, these can only be guidelines. Specific courses and groups of students vary greatly. Teachers must modify schedules provided in instructional materials by considering such factors as the length of the course, the length of class period, the students' abilities and backgrounds, their own teaching styles, and the laboratory activities to be used. It is important, however, to present a complete view of the content area. If you do not, students will receive a narrow, incomplete concept of technology. For example, emphasizing the design section, while leaving out the context sections exploring various technological systems, leads students to believe design is the most important aspect of technology. Likewise, failing to address the positive and negative impacts of technology leaves students poorly prepared to assess the value of technological advancements. Some proposed schedules for 9-week, 12-week, 18-week, and 36-week Introduction to Technology programs are presented in **Figure IN-4.**

Working with the Instructional System

This Introduction to Technology program is an instructional system with several components. These components are the teacher, students, textbook, Tech Lab Workbook, and Teacher's Manual. Each component plays a unique role in helping students develop a level of technological literacy and competence. Later, students can use their knowledge in adult citizenship, career, family, and consumer roles.

Role of the Teacher

Teachers and curriculum developers generally have several instructional systems available to meet their needs. Some systems try to provide all the answers. They are designed to be "teacherproof." Textbook writers carefully plan and orchestrate each class period. The teacher becomes an actor, who is expected to deliver predetermined lines from the script and conduct prescribed activities. This type of material has certain attractions. The material is predictable and easy to follow. This approach, however, diminishes the professional role of the teacher and is difficult to modify when it does not meet the needs of specific groups of students. By contrast, the system on which this program was developed places the teacher in the key role in the educational process. This system assigns the teacher at least three distinct roles:

- ❑ **Curriculum planner.** The textbook, Tech Lab Workbook, and Teacher's Manual provide suggested objectives, content, course outlines, study questions, and student activities. These suggested elements are recommendations. They are general guides that can be modified to fit the local situation. All modifications, however, should maintain the program's overall view of technology.
- ❑ **Technical expert.** The program presents students with a conceptual view of technology and technology's impact on their ways of life. Only through hands-on investigation, practice, and exploration can they develop a truly meaningful understanding of this vital force in their lives. The textbook and Tech Lab Workbook provide many appropriate activities for the class. The presentation of the safe and appropriate techniques to complete these activities is the teacher's responsibility.
- ❑ **Manager of education.** Teaching technology education has moved a long way from the "tell them, show them, test them" mode of the past. Technology education now challenges students to become actively involved in designing; problem solving; interpreting; assessing; and other vital, higher-level activities. This movement makes the teacher the manager of education. In this role, teachers plan, organize, actuate, and control. They should do all of the following:
    1. Plan the program. Establish the major goals of the course and the time schedule for the class.
    2. Organize the learning. Adopt structures and procedures for carrying out the various learning activities.
    3. Actuate the planned activities. Guide and motivate students as they plan and participate in specific learning events.
    4. Control the learning process. Evaluate student achievement, your effectiveness, and program adequacy in light of the program goals.

<table>
<tr><th colspan="5">Some Proposed Course Schedules</th></tr>
<tr><th rowspan="2">Chapter Title and Number</th><th colspan="4">Length of the Course in Weeks</th></tr>
<tr><th>9</th><th>12</th><th>18</th><th>36</th></tr>
<tr><td colspan="5">Section 1 - Scope of Technology</td></tr>
<tr><td>1. What Is Technology?
2. Technology as a System
3. Contexts of Technology</td><td>1 week</td><td>1 week</td><td>2 weeks</td><td>2 weeks</td></tr>
<tr><td colspan="5">Section 2 - Resources and Technology</td></tr>
<tr><td>4. Tools and Technology
5. Materials and Technology</td><td>1 week</td><td>1 week</td><td>2 weeks</td><td>2 weeks</td></tr>
<tr><td>6. Energy and Technology
7. Information and Technology
8. People, Time, Money, and Technology</td><td>1 week</td><td>1 week</td><td>1 week</td><td>2 weeks</td></tr>
<tr><td colspan="5">Section 3 - Creating Technology</td></tr>
<tr><td>9. Invention and Innovation
10. The Design Process
11. Identifying Problems
12. Researching Problems
13. Creating Solutions
14. Selecting and Refining Solutions
15. Modeling Solutions
16. Testing Solutions
17. Communicating Solutions
18. Improving Solutions</td><td>2 weeks</td><td>4 weeks</td><td>6 weeks</td><td>10 weeks</td></tr>
<tr><td colspan="5">Section 4 - Technological Contexts</td></tr>
<tr><td>19. Agricultural and Related Technology</td><td rowspan="3">1 week</td><td rowspan="2">1 week</td><td rowspan="2">2 weeks</td><td>2 weeks</td></tr>
<tr><td>20. Construction Technology</td><td>3 weeks</td></tr>
<tr><td>21. Energy-Conversion Technology</td><td rowspan="3">2 weeks</td><td rowspan="3">2 weeks</td><td>2 weeks</td></tr>
<tr><td>22. Information and Communication Technology</td><td rowspan="2">1 week</td><td>3 weeks</td></tr>
<tr><td>23. Manufacturing Technology</td><td>3 weeks</td></tr>
<tr><td>24. Medical Technology</td><td rowspan="2">1 week</td><td rowspan="2">1 week</td><td rowspan="2">2 weeks</td><td>2 weeks</td></tr>
<tr><td>25. Transportation Technology</td><td>3 weeks</td></tr>
<tr><td colspan="5">Section 5 - Technology and Society</td></tr>
<tr><td>26. Technological Impacts
27. Technology and the Future</td><td>1 week</td><td>1 week</td><td>1 week</td><td>2 weeks</td></tr>
</table>

Figure IN-4

Role of the Students

Society builds schools and hires teachers so students can learn important information and develop critical skills. This culture wants to equip its young members to meet the challenges of citizenship. Therefore, students must be active players in the learning process. Technology education must challenge students to do the following:

- ❑ Participate in the class activities.
- ❑ Work up to their potentials.
- ❑ Develop cooperative attitudes.
- ❑ Explore their creative, mental, and physical abilities.

Opportunities to engage in meaningful design and problem-solving activities and to make significant choices in the class are an essential part of the process. Laboratory activities are one way to make this happen. The text and Tech Lab Workbook contain suggested activities that make the content more meaningful. These activities, however, can be replaced as student interest and local conditions dictate. *Caution:* Check carefully that each activity makes the content of the course more meaningful, while helping to develop student technical abilities. Only use activities supporting the goals of the course and specific units of study. There is no reason to select activities simply because the students like them, the school needs the product, plans and procedure sheets are available, or they have been used before. Also, select activities providing a balance between individual and group endeavors. Students need to develop both individual skills (goal setting, good work habits, and creative thinking) and group competence (leadership, followership, cooperation, and responsibility to others).

Accommodating Individual Differences

Each student has a unique set of abilities and a particular learning style. Education has often tried to teach students as if a single set of strategies and a specific attainment goal can accommodate the needs of all students. This approach has proven to be less than fruitful. The modern educational challenge is to meet overall educational goals, while addressing the vast range of individual differences present in most classrooms. Technology education has the power to meet this challenge, if it is approached through a design and problem-solving approach, where students work in groups to meet stated objectives. This approach requires the teacher to accept several philosophical tenets:

- ❑ Understanding the process of developing, operating, and using technology is more important than learning specific facts about technology.
- ❑ The study of technology is focused on the why of technology, rather than on how to use specific tools or machines.
- ❑ Technology education contributes to a society requiring group participation a great deal more than individual, competitive action.
- ❑ Society needs and can accommodate people with diverse interests and abilities and, therefore, does not require people with uniform abilities, goals, or knowledge.
- ❑ Society also needs people who understand technology and can help design our technological future through citizen, consumer, and career roles.
- ❑ Members of society should develop habits enabling them to contribute their best efforts within the constraints of time and resources.

Developing programs having this vision requires a unique view of students and subject matter content. This vision requires instruction to be conceptual rather than fact focused, group oriented rather than individual oriented, and flexible rather than rigid in its approach. Instruction should be structured so students who apply themselves can taste success. This means there must be opportunities for students to develop and use their specific talents. The chart in **Figure IN-5** lists some of the ways this instructional system creates these opportunities as students explore technology.

Helping Students Gain a Basic Knowledge of Technology	
Textbooks	Films
Filmstrips	Video recordings
Slide series	Field trips
Guest speakers	Research

Helping Students Develop Problem-Solving Abilities	
Product and structure design	System development
Process planning	Material testing

Helping Students Develop Creativity	
Product and structure design	Brainstorming
Communication-message design	Advertising design
Package design	Vehicle design

Helping Students Apply Technology	
Product and structure design	Constructing structures
Manufacturing products	Designing messages
Energy-system design	Material testing
Vehicle design	

Helping Students Assess Technology	
Product-design review	Material testing
Energy-system analysis	Research
Environmental-impact analysis	Future projections

Helping Students Develop Communications Skills	
Class discussions	Group interaction
Drawing and sketching	Product-design presentation
Laboratory summaries	Written and oral reports

Helping Students Develop Leadership Skills	
Decision making	Goal setting
Goal assessment	Managing activities
Cooperative group activities	

Figure IN-5

Role of the Textbook

An ever-changing world challenges schools to remain relevant. Citizens are expected to grow intellectually throughout their lives. This requires a strong foundation in the basics: sciences, technologies, and humanities. Moreover, individuals must develop the skills needed to seek out new information. This development rests on the ability to read. The textbook plays a vital role in these challenges. This text presents students with the rudiments of technology and encourages them to develop the ability to comprehend what they read. Each student should, then, have a copy of the course text. Students should be expected to read the text and be encouraged to take it home. Class time is not for reading the textbook. Rather, it is the time for technology to come alive in laboratory activities, presentations and discussions, films and videos, and other interactive processes. *Exploring Design, Technology, & Engineering* opens the world of technology to all students. This text is colorfully illustrated and interestingly written. The textbook is the source of information for class discussions and activity development.

Role of the Tech Lab Workbook

The Tech Lab Workbook gives students guidance and support as they engage in their study of technology. The workbook is a consumable including the following features:

- ❑ Laboratory activity sheets for selected activities included in the textbook.
- ❑ Alternative, problem-centered activities for most chapters in the textbook.
- ❑ Study questions to support the textbook.

The teacher can conduct the course without the Tech Lab Workbook. To do so, however, requires more work and preparation time. This workbook is, in every respect, a guide for completing the mental and physical activities envisioned for the class.

Role of the Teacher's Manual

The Teacher's Manual goes well beyond an answer key for the textbook and Tech Lab Workbook study questions. This resource also does the following:

- ❑ Offers a rationale for the program that can be used to inform students, parents, guidance personnel, and administrators of the value of the study of technology.
- ❑ Suggests approaches and procedures for developing and presenting each section, including the following elements:
    1. An introduction.
    2. Objectives.
    3. Suggested length and schedule.
    4. Teacher preparatory activities.
    5. Suggestions for managing student activities.
- ❑ Presents ways of selecting and presenting laboratory activities.

The Teacher's Manual is not a script. This resource is a course-development guide and, therefore, should be read in full before the course is started. The does not attempt to provide a cookbook approach to each activity, with the ingredients and procedures carefully prescribed. Nor have the authors developed the program for an ideal or average class, with the expectation that all other classes will, somehow, fit this mythical mold. You, as the primary educational agent in the classroom, should read and study the guide to do the following:

- ❑ Develop an overall view of the goals, content, and activities for the class.
- ❑ Prepare long-range plans, including the need for content-presentation development, media selection and development, activity preparation and modification, equipment and supply purchases, and professional growth.
- ❑ Schedule daily learning activities.
- ❑ Prepare daily lesson and activity plans.
- ❑ Design an evaluation system.
- ❑ Plan a classroom-management schedule.

Sections as Content Organizers

In many school subjects, content is presented through a series of discrete topics that are generally the themes of specific chapters. This approach causes students to perceive the course as a group of isolated concepts and series of facts. A more useful approach is to organize content into larger sections. This is the approach suggested for this instructional system. Each section or unit of study has an overall uniting theme integrating a number of chapters and activities. Using this approach, students will more likely see technology as a unified and purposeful human activity. Each component of the total instructional systems must be placed in its proper context, if learning is to be maximized. No single element can do the whole job. All the information cannot be presented. The student cannot solve all the problems. The textbook cannot provide all the content, nor make it concrete. The laboratory activities cannot be the reason for the class. If each component is used to do what it can do well, however, the class will be challenging, fun, and informative. That's your challenge.

Standards for Technological Literacy Correlation Chart

The International Technology and Engineering Educators Association (ITEEA) and its Technology for All Americans Project developed *Standards for Technological Literacy: Content for the Study of Technology* to identify the essential core of technological knowledge and skills for students in grades K–12. This work defines 20 separate standards, divided into five broad categories. Within each standard, benchmark topics are defined for four different grade levels:

- ❑ Grades K–2.
- ❑ Grades 3–5.
- ❑ Grades 6–8.
- ❑ Grades 9–12.

The following chart lists the standards and the benchmark topics for grades 6–8. Adjacent to each standard and benchmark topic are the chapter and page references identifying material in *Exploring Design, Technology, & Engineering* relating to the item.

1. Students will develop an understanding of the characteristics and scope of technology.	Chapter 1: pages 25–28, 30–31 Chapter 3: pages 59–67 Chapter 9: pages 210–212, 216–217 Chapter 23: pages 566–567
New products and systems can be developed to solve problems or to help do things that could not be done without the help of technology.	Chapter 1: pages 25–28 Chapter 9: page 210
The development of technology is a human activity and is the result of individual or collective needs and the ability to be creative.	Chapter 1: pages 26–28 Chapter 3: pages 59–67 Chapter 9: pages 210–212
Technology is closely linked to creativity, which has resulted in innovation.	Chapter 3: pages 59–67 Chapter 9: pages 216–217
Corporations can often create demand for a product by bringing it onto the market and advertising it.	Chapter 1: pages 30–31 Chapter 23: pages 566–567
2. Students will develop an understanding of the core concepts of technology.	Chapter 1: pages 25–32 Chapter 2: pages 37–40, 44–48, 50–51 Chapter 10: pages 225–228 Chapter 11: pages 254–257 Chapter 14: pages 302–303 *(continued)*

2. Students will develop an understanding of the core concepts of technology *(continued)*.	Chapter 16: pages 360–361 Chapter 18: pages 396–400
Technological systems include input, processes, output, and, at times, feedback.	Chapter 2: pages 38–40
Systems thinking involves considering how every part relates to others.	Chapter 2: pages 37–38
An open-loop system has no feedback path and requires human intervention, while a closed-loop system uses feedback.	Chapter 2: pages 50–51
Technological systems can be connected to one another.	Chapter 2: pages 37–38
Malfunctions of any part of a system may affect the function and quality of the system.	Chapter 2: pages 38–40
Requirements are the parameters placed on the development of a product or system.	Chapter 11: pages 254–257
Trade-off is a decision process recognizing the need for careful compromises among competing factors.	Chapter 16: pages 360–361
Different technologies involve different sets of processes.	Chapter 2: pages 44–48
Maintenance is the process of inspecting and servicing a product or system on a regular basis in order for it to continue functioning properly, to extend its life, or to upgrade its capability.	Chapter 2: pages 38–40
Controls are mechanisms or particular steps that people perform using information about the system that causes systems to change.	Chapter 2: pages 50–51
3. Students will develop an understanding of the relationships among technologies and the connections between technology and other fields of study.	Chapter 1: page 32 Chapter 2: pages 37–38, 52 Chapter 3: pages 72–74 Chapter 4: page 99 Chapter 5: page 129 Chapter 6: page 154 Chapter 7: pages 165–167, 176 Chapter 8: page 199 Chapter 9: page 218 Chapter 10: page 239 Chapter 11: page 258 Chapter 12: page 274 Chapter 13: page 297 Chapter 14: page 316 Chapter 15: page 343 Chapter 16: page 361 Chapter 17: pages 388–389 Chapter 18: page 405 *(continued)*

3. Students will develop an understanding of the relationships among technologies and the connections between technology and other fields of study *(continued)*.	Chapter 19: page 438 Chapter 20: page 463 Chapter 21: page 493 Chapter 22: page 535 Chapter 23: page 568 Chapter 24: page 594 Chapter 25: page 619 Chapter 26: page 650 Chapter 27: page 668
Technological systems often interact with one another.	Chapter 2: pages 37–38
A product, system, or environment developed for one setting may be applied to another setting.	Chapter 3: pages 72–73
Knowledge gained from other fields of study has a direct effect on the development of technological products and systems.	Chapter 7: pages 165–167
4. Students will develop an understanding of the cultural, social, economic, and political effects of technology.	Chapter 1: pages 25–26 Chapter 3: pages 69–71 Chapter 8: pages 183–193 Chapter 26: pages 639–648
The use of technology affects humans in various ways, including their safety, comfort, choices, and attitudes about technology's development and use.	Chapter 3: pages 69–70 Chapter 8: pages 183–185 Chapter 26: pages 639–648
Technology, by itself, is neither good nor bad, but decisions about the use of products and systems can result in desirable or undesirable consequences.	Chapter 3: pages 70–71 Chapter 26: pages 639–648
The development and use of technology poses ethical issues.	Chapter 26: pages 639–648
Economic, political, and cultural issues are influenced by the development and use of technology.	Chapter 8: pages 183–193 Chapter 26: pages 639–648
5. Students will develop an understanding of the effects of technology on the environment.	Chapter 2: pages 48–50 Chapter 26: pages 642–644 Chapter 27: pages 663–667
The management of waste produced by technological systems is an important societal issue.	Chapter 2: pages 48–50 Chapter 26: pages 642–667
Technologies can be used to repair damage caused by natural disasters and to break down waste from the use of various products and systems.	Chapter 26: pages 642–644
Decisions to develop and use technologies often put environmental and economic concerns in direct competition with one another.	Chapter 26: pages 642–644
6. Students will develop an understanding of the role of society in the development and use of technology.	Chapter 2: pages 40–42 Chapter 3: pages 68–71 *(continued)*

6. Students will develop an understanding of the role of society in the development and use of technology *(continued)*.	Chapter 8: pages 183–185 Chapter 9: pages 206–208, 213–214 Chapter 26: pages 629–639
Throughout history, new technologies have resulted from the demands, values, and interests of individuals, businesses, industries, and societies.	Chapter 2: page 42 Chapter 3: pages 68–71 Chapter 9: pages 206–208
The use of inventions and innovations has led to changes in society and the creation of new needs and wants.	Chapter 2: pages 40–41 Chapter 3: pages 69–70 Chapter 9: pages 213–214
Social and cultural priorities and values are reflected in technological devices.	Chapter 3: pages 68–71 Chapter 26: pages 629–639
Meeting societal expectations is the driving force behind the acceptance and use of products and systems.	Chapter 3: pages 68–71
7. Students will develop an understanding of the influence of technology on history.	Chapter 1: pages 26–28 Chapter 2: page 48 Chapter 3: pages 60–61, 72–73 Chapter 4: pages 82–85 Chapter 9: pages 206–208 Chapter 10: page 239 Chapter 20: pages 443–444
Many inventions and innovations have evolved by using slow and methodical processes of tests and refinements.	Chapter 9: pages 206–208
The specialization of function has been at the heart of many technological improvements.	Chapter 3: pages 72–73
The design and construction of structures for service or convenience have evolved from the development of techniques for measurement, controlling systems, and the understanding of spatial relationships.	Chapter 2: page 48 Chapter 4: pages 82–85 Chapter 20: pages 443–444
In the past, an invention or innovation was not usually developed with the knowledge of science.	Chapter 3: pages 60–61
8. Students will develop an understanding of the attributes of design.	Chapter 10: pages 225–226, 230–231, 237–238 Chapter 11: pages 252–256 Chapter 12: pages 267–271 Chapter 13: pages 283–284 Chapter 14: pages 303–308, 310–314 Chapter 15: pages 340–342 Chapter 16: pages 349–355 Chapter 17: page 367 Chapter 18: pages 396–400
Design is a creative planning process that leads to useful products and systems.	Chapter 10: pages 225–226

There is no perfect design.	Chapter 10: pages 237–238
Requirements for a design are made up of criteria and constraints.	Chapter 10: pages 230–231 Chapter 11: pages 252–256
9. Students will develop an understanding of engineering design.	Chapter 10: pages 228–238 Chapter 11: pages 250–252 Chapter 12: pages 267–273 Chapter 13: pages 277–296 Chapter 14: pages 302–303, 310–314 Chapter 15: pages 320–322, 340–342 Chapter 16: pages 347–348, 358–361 Chapter 17: pages 386–387 Chapter 18: pages 404–405
Design involves a set of steps, which can be performed in different sequences and repeated as needed.	Chapter 10: pages 228–238 Chapter 11: pages 250–252 Chapter 12: pages 267–273 Chapter 13: pages 277–296 Chapter 14: pages 302–303, 310–314 Chapter 15: pages 340–342 Chapter 16: pages 358–360 Chapter 17: pages 386–387 Chapter 18: pages 404–405
Brainstorming is a group problem-solving process in which each person in the group presents his or her ideas in an open forum.	Chapter 10: page 233 Chapter 13: page 280
Modeling, testing, evaluating, and modifying are used to transform ideas into practical solutions.	Chapter 10: pages 233–238 Chapter 14: pages 310–314 Chapter 15: pages 320–322 Chapter 16: pages 347–348, 360–361 Chapter 18: pages 404–405
10. Students will develop an understanding of the role of troubleshooting, research and development, invention and innovation, and experimentation in problem solving.	Chapter 9: pages 205–206, 210, 216–218 Chapter 12: pages 265–267 Chapter 13: page 297 Chapter 16: pages 355–358
Troubleshooting is a problem-solving method used to identify the cause of a malfunction in a technological system.	Chapter 9: page 210
Invention is a process of turning ideas and imagination into devices and systems. Innovation is the process of modifying an existing product or system to improve it.	Chapter 9: pages 205–206, 216–218
Some technological problems are best solved through experimentation.	Chapter 9: page 210 Chapter 16: pages 355–358

11. Students will develop abilities to apply the design process.	Chapter 10: pages 228–238 Chapter 11: pages 252–257 Chapter 12: page 274 Chapter 13: pages 283–296 Chapter 14: pages 310–315 Chapter 15: pages 340–342 Chapter 16: pages 348–349 Chapter 17: pages 388–389 Chapter 18: pages 396–400
Apply a design process to solve problems in and beyond the laboratory-classroom.	Chapter 10: pages 228–238
Specify criteria and constraints for the design.	Chapter 11: pages 252–257
Make two-dimensional and three-dimensional representations of the designed solution.	Chapter 10: pages 235–236 Chapter 13: pages 283–296 Chapter 14: pages 314–315 Chapter 15: pages 340–342
Test and evaluate the design in relation to pre-established requirements, such as criteria and constraints, and refine as needed.	Chapter 10: page 236 Chapter 11: pages 252–256 Chapter 14: pages 310–315 Chapter 16: pages 348–349 Chapter 18: pages 396–400
Make a product or system and document the solution.	Chapter 10: pages 236–237 Chapter 17: pages 388–389
12. Students will develop the abilities to use and maintain technological products and systems.	Chapter 2: pages 44–48 Chapter 4: pages 81–82 Chapter 4 Activities: pages 101–107 Chapter 5: page 129 Chapter 5 Activity: pages 131–135 Chapter 6: page 154 Chapter 6 Activities: pages 156–163 Chapter 7: page 176 Chapter 7 Activity: pages 177–181 Chapter 8: page 199 Chapter 9 Activity: pages 220–222 Chapter 10: page 239 Chapter 10 Activity: pages 241–243 Chapter 11: page 258 Chapter 12: pages 262–264, 274 Chapter 15: pages 330–335, 343 Chapter 18: page 405 Chapter 18 Activity: pages 407–409 *(continued)*

12. Students will develop the abilities to use and maintain technological products and systems *(continued)*.	Chapter 19: page 438 Chapter 19 Activity: page 440 Chapter 20 Activity: pages 465–470 Chapter 21: page 494 Chapter 21 Activity: pages 495–496 Chapter 22: page 536 Chapter 22 Activity: pages 538–540 Chapter 23 Activity: pages 570–573 Chapter 24 Activity: pages 596–597 Chapter 25 Activity: pages 621–625 Chapter 26: page 650
Use information provided in manuals, protocols, or by experienced people to see and understand how things work.	Chapter 7: page 176 Chapter 10: page 239 Chapter 11: page 258 Chapter 12: pages 262–264 Chapter 15: page 343 Chapter 18: page 405 Chapter 19: page 438 Chapter 22: page 536 Chapter 26: page 650
Use tools, materials, and machines safely to diagnose, adjust, and repair systems.	Chapter 4: pages 81–92 Chapter 15: pages 330–335
Use computers and calculators in various applications.	Chapter 5: page 129 Chapter 6: page 154 Chapter 8: page 199 Chapter 12: page 274 Chapter 19: page 438 Chapter 21: page 494 Chapter 22: page 536 Chapter 26: page 650
Operate and maintain systems in order to achieve a given purpose.	Chapter 2: pages 44–48
13. Students will develop the abilities to assess the impact of products and systems.	Chapter 4 Activities: pages 101–107 Chapter 5 Activity: pages 131–135 Chapter 6 Activities: pages 156–163 Chapter 7 Activity: pages 177–181 Chapter 10: page 236 Chapter 12: pages 267–274 Chapter 19: page 438 Chapter 19 Activity: page 440 *(continued)*

13. Students will develop the abilities to assess the impact of products and systems *(continued).*	Chapter 20: page 463 Chapter 20 Activity: pages 465–470 Chapter 21 Activity: pages 495–496 Chapter 22 Activity: pages 538–540 Chapter 23 Activity: pages 570–573 Chapter 24: page 594 Chapter 24 Activity: pages 596–597 Chapter 25 Activity: pages 621–625 Chapter 26: pages 639–648, 650 Chapter 26 Activity: pages 652–653 Chapter 27: page 660
Design and use instruments to gather data.	Chapter 12: pages 267–271, 274 Chapter 19: page 438 Chapter 20: page 463 Chapter 24: page 594 Chapter 26: page 650
Use data collected to analyze and interpret trends in order to identify the positive or negative effects of a technology.	Chapter 26: pages 639–648, 650
Identify trends and monitor potential consequences of technological development.	Chapter 26: pages 639–648 Chapter 27: page 660
Interpret and evaluate the accuracy of the information obtained and determine if it is useful.	Chapter 12: pages 272–273
14. Students will develop an understanding of and be able to select and use medical technologies.	Chapter 19: pages 429–430 Chapter 24: pages 577–578, 583–584, 589–594
Advances and innovations in medical technologies are used to improve healthcare.	Chapter 24: pages 577–578
Sanitation processes used in the disposal of medical products help to protect people from harmful organisms and disease, and shape the ethics of medical safety.	Chapter 24: pages 589–592
The vaccines developed for use in immunization require specialized technologies to support environments in which a sufficient amount of vaccines are produced.	Chapter 24: pages 583–584
Genetic engineering involves modifying the structure of DNA to produce novel genetic make-ups.	Chapter 19: pages 429–430 Chapter 24: pages 592–594
15. Students will develop an understanding of and be able to select and use agricultural and related biotechnologies.	Chapter 19: pages 417–437
Technological advances in agriculture directly affect the time and number of people required to produce food for a large population.	Chapter 19: pages 417–425

A wide range of specialized equipment and practices is used to improve the production of food, fiber, fuel, and other useful products and in the care of animals.	Chapter 19: pages 417–426
Biotechnology applies the principles of biology to create commercial products or processes.	Chapter 19: pages 429–430
Artificial ecosystems are human-made complexes that replicate some aspects of the natural environment.	Chapter 19: pages 435–436
The development of refrigeration, freezing, dehydration, preservation, and irradiation provide long-term storage of food and reduce the health risks caused by tainted food.	Chapter 19: pages 430–435
16. Students will develop an understanding of and be able to select and use energy and power technologies.	Chapter 6: pages 137–140 Chapter 21: pages 476–490, 492–493 Chapter 26: pages 640–642 Chapter 27: pages 664–665
Energy is the capacity to do work.	Chapter 6: pages 139–140 Chapter 21: page 476
Energy can be used to do work, using many processes.	Chapter 6: pages 137–138 Chapter 21: pages 476–490
Power is the rate at which energy is converted from one form to another or transferred from one place to another, or the rate at which work is done.	Chapter 6: page 140
Power systems are used to drive and provide propulsion to other technological products and systems.	Chapter 21: pages 476–490
Much of the energy used in our environment is not used efficiently.	Chapter 21: pages 492–493 Chapter 26: pages 640–642 Chapter 27: pages 664–665
17. Students will develop an understanding of and be able to select and use information and communication technologies.	Chapter 7: page 172 Chapter 17: pages 370–382 Chapter 22: pages 503, 507–526
Information and communication systems allow information to be transferred from human to human, human to machine, and machine to human.	Chapter 22: pages 507–522
Communication systems are made up of a source, encoder, transmitter, receiver, decoder, and destination.	Chapter 22: pages 514, 520–524
The design of a message is influenced by such factors as the intended audience, medium, purpose, and nature of the message.	Chapter 22: pages 524–526
The use of symbols, measurements, and drawings promotes clear communication by providing a common language to express ideas.	Chapter 17: pages 370–382 Chapter 22: pages 503, 518
18. Students will develop an understanding of and be able to select and use transportation technologies.	Chapter 25: pages 600–601, 603, 610–618

Transporting people and goods involves a combination of individuals and vehicles.	Chapter 25: pages 600–601
Transportation vehicles are made up of subsystems, such as structural, propulsion, suspension, guidance, control, and support, that must function together for a system to work effectively.	Chapter 25: pages 610–618
Governmental regulations often influence the design and operation of transportation systems.	Chapter 25: pages 616–618
Processes, such as receiving, holding, storing, loading, moving, unloading, delivering, evaluating, marketing, managing, communicating, and using conventions are necessary for the entire transportation system to operate efficiently.	Chapter 25: page 603
19. Students will develop an understanding of and be able to select and use manufacturing technologies.	Chapter 5: pages 115–120, 122 Chapter 23: pages 545–557, 561–567
Manufacturing systems use mechanical processes that change the form of materials through the processes of separating, forming, combining, and conditioning them.	Chapter 23: pages 545–557
Manufactured products may be classified as durable and non-durable.	Chapter 23: pages 545–557
The manufacturing process includes the designing, development, making, and servicing of products and systems.	Chapter 23: pages 545–557
Chemical technologies are used to modify or alter chemical substances.	Chapter 5: page 122
Materials must first be located before they can be extracted from the earth through such processes as harvesting, drilling, and mining.	Chapter 5: pages 115–120
Marketing a product involves informing the public about it as well as assisting in selling and distributing it.	Chapter 23: pages 566–567
20. Students will develop an understanding of and be able to select and use construction technologies.	Chapter 20: pages 450–454, 458–459
The selection of designs for structures is based on factors such as building laws and codes, style, convenience, cost, climate, and function.	Chapter 20: pages 450–454, 458–459
Structures rest on a foundation.	Chapter 20: pages 453–454
Some structures are temporary, while others are permanent.	Chapter 20: pages 450–453, 458
Buildings generally contain a variety of subsystems.	Chapter 20: pages 458–459

Basic Skills Chart

The Basic Skills Chart has been designed to identify those activities in the *Exploring Design, Technology, & Engineering* text and Tech Lab Workbook specifically encouraging the development of basic skills. Academic areas addressed in the chart include reading, writing, verbal (other than reading and writing), math, science, and analytical.

- **Reading** activities include assignments designed to improve comprehension of information presented in the chapter. Some are designed to improve understanding of vocabulary terms.
- **Writing** activities allow students to practice composition skills, such as letter writing, informative writing, and creative writing.
- **Verbal** activities encourage students to organize ideas, develop interpersonal and group speaking skills, and respond appropriately to verbal messages. Activities include oral reports and interviews.
- **Math** activities require students to use basic principles of math, as well as computation skills, to solve typical problems.
- **Science** activities call for students to use fundamental principles of science to solve typical problems.
- **Analytical** activities involve the higher-order skills needed for thinking creatively, making decisions, solving problems, visualizing information, reasoning, and knowing how to learn.

Activities are broken down by chapter, and a page number is given to locate the activity.

Section 1	
Reading	**Text:** **1:** Technology and History (26); Technology and Knowledge (28); Technology and Action (30); All Subjects Connection (32) **2:** System Goals (40); System Inputs (41); System Processes (44); Container Shipping (46); System Outputs (48); System Feedback (50) **3:** Historical Context (59); Peanut Butter (62); Societal Context (68); Functional Context (72)
Writing	
Verbal	**Text:** **1:** Technology and Language (31)
Math	**Text:** **3:** Mathematics Connection (74)
Science	**Text:** **1:** Science Connection (32); Activity 1 (33) **2:** Science Connection (52) **3:** Science Connection (74)

<table>
<tr><td>Analytical</td><td>Text:
1: All Subjects Connection (32); Social Studies Connection (32); Activity 2 (33); Activity 3 (33)
2: All Subjects Connection (52); Social Studies Connection (52); Activity 1 (52); Activity 2 (52); Activity 3 (52)
3: Social Studies Connections (74); Activity 1 (74); Activity 2 (74); Activity 3 (74)
Chapter 1 Activity A: What Is Technology? (35)
Chapter 2 Activity A: Using Technology to Make a Product (53)</td></tr>
<tr><td></td><td>Tech Lab Workbook:
Chapter 2 Activity A Worksheet: Using Technology to Make a Product (9)
2-1: Marble Counter (13)
2-2: Technological Systems (15)</td></tr>
<tr><td colspan="2">Section 2</td></tr>
<tr><td>Reading</td><td>Text:
4: Toolmakers (80); Tools (81); Mechanisms (89); Machines (94)
5: Types of Materials (111); Engineering Materials (112); Obtaining Material Resources (115); Processing Raw Materials (120)
7: Data, Information, and Knowledge (168); Technological Knowledge (171); The Information Age (172); Gaining Information and Knowledge (172); Social Studies Connection (176); Activity 1 (176); Activity 2 (176)
8: Work and a Changing Society (184); Jobs and Technology (185); Flight Controls (187); Jobs in the Information Age (188); Jobs in Technology (189); People and Jobs (191); Advancing in a Career (195); Time and Technology (196)</td></tr>
<tr><td>Writing</td><td></td></tr>
<tr><td>Verbal</td><td>Text:
6: Language Arts Connection (154)
7: Social Studies Connection (176); Activity 2 (176)</td></tr>
<tr><td>Math</td><td>Text:
4: Mathematics Connection (99)
5: Mathematics Connection (129)
6: Mathematics Connection (154)
7: Mathematics Connection (176)
8: Mathematics Connection (199)</td></tr>
<tr><td>Science</td><td>Text:
4: Lasers (84); Science Connection (99)
5: Geothermal Energy (117); Properties of Materials (123); Science Connection (129); Activity 2 (129); Activity 3 (129)
6: Energy and Work (139); Energy and Power (140); The Nature of Energy (142); Types of Energy Sources (143); Common Energy Supplies (144); Biogas (147); Alternate Energy Sources (151); Social Studies Connection (154); Science Connection (154); Activity 1 (155); Activity 2 (155); Activity 3 (155)
(continued)</td></tr>
</table>

Science *(continued)*	**Text:** **7:** Scientific Knowledge (170); Science Connection (176) **8:** Money and Technology (198); Science Connection (199) **Chapter 6 Activity A:** Energy as a Resource (156) **Chapter 6 Activity B:** Energy and Machines as Resources (163)
	Tech Lab Workbook: **6-1:** Tools and Energy as Resources (39) **Chapter 5 Activity A Worksheet:** Materials as a Resource (31) **Chapter 6 Activity A Worksheet:** Energy as a Resource (37)
Analytical	**Text:** **4:** Social Studies Connection (99); Activity 1 (99); Activity 2 (100); Activity 3 (100) **5:** Social Studies Connection (129); Activity 1 (129) **6:** Social Studies Connection (154) **7:** Social Studies Connection (176); Activity 1 (176) **8:** Social Studies Connection (199); Activity 1 (199); Activity 2 (199); Activity 3 (199) **Chapter 4 Activity A:** Tools and Materials as Resources (101) **Chapter 4 Activity B:** Tools as a Resource (107) **Chapter 5 Activity A:** Materials as a Resource (131) **Chapter 7 Activity A:** Information as a Resource (177)
	Tech Lab Workbook: **Chapter 4 Activity A Worksheet:** Tools and Materials as Resources (21) **5-1:** Product Analysis (33) **Chapter 7 Activity A Worksheet:** Information as a Resource (43) **7-1:** Developing an Information Chart (45)
Section 3	
Reading	**Text:** **9:** Invention (206); Discovery vs Invention (208); The Invention Process (210); Inventors (212); Reasons for Inventing (213); Invention Teams (214); High-Rise Buildings (215); Innovation (216) **10:** What Is Design? (225); Types of Design (226); Steps of the Design Process (228) **11:** Needs and Wants (247); Problems and Opportunities (248); Types of Problems (249); Problem Statements (252); Criteria (252); Constraints (254); Design Briefs (256) **12:** Types of Research (262); The Research Process (267); Language Arts Connection (274) **13:** Ideation (277); Sketching (283); Elegant Solutions (286) **14:** Selecting Solutions (302); Aspects of Design (303); Design Sheets (308); Evaluation Grids (308); Refining Solutions (310); Renderings (314); Detailed Sketches (314) **15:** What Is a Model? (320); Why Are Models Used? (320); Types of Models (322); Modeling Materials (329); Modeling Tools (339); Making a Model (340) *(continued)*

Reading *(continued)*	**16:** Using the Design Criteria (348); Testing for Certain Characteristics (349); Testing Environments (355); The Testing Procedure (358) **17:** Communication Documents (370); Drawings (370); Types of Working Drawings (374); Drawing Classifications (376); Mechanical Drawing Tools (382); Making Mechanical Drawings (386) **18:** Reasons for Improvement (396); Sources of Improvement (400); Making Improvements (404); Activity 1 (405)
Writing	**Text:** **9:** Language Arts Connection (218) **10:** Language Arts Connection (239) **11:** Language Arts Connection (258); Activity 3 (258) **13:** Language Arts Connection (297) **15:** Activity 2 (343) **16:** Language Arts Connection (361) **17:** Language Arts Connection (389); Activity 1 (389); Activity 3 (389) **18:** Language Arts Connection (405)
	Tech Lab Workbook: **Chapter 18 Activity A Worksheet:** Communication Design (95)
Verbal	**Text:** **16:** Activity 2 (361) **17:** Documenting and Presenting the Solution (387); Activity 3 (389) **18:** Activity 3 (405) **Chapter 18 Activity A:** Communication Design (407)
	Tech Lab Workbook: **Chapter 18 Activity A Worksheet:** Communication Design (95)
Math	**Text:** **12:** Mathematics Connection (274); Activity 2 (274) **13:** Mathematics Connection (297) **14:** Mathematics Connection (316) **15:** Mathematics Connection (343) **17:** Mathematics Connection (388)
Science	**Text:** **9:** Science Connection (218) **10:** Science Connection (239) **11:** Science Connection (258) **12:** Science Connection (274) **13:** Science Connection (297) **14:** Science Connection (316) **15:** Science Connection (343) **16:** Science Connection (361) *(continued)*

Science *(continued)*	**Text:** **17:** Science Connection (388) **18:** Science Connection (405)
Analytical	**Text:** **9:** Social Studies Connection (218); Activity 1 (218); Activity 2 (218); Activity 3 (218); Invention and Innovation (220) **10:** Social Studies Connection (239); Activity 1 (239); Activity 2 (239); Activity 3 (239) **11:** Recognizing Problems (250); Social Studies Connection (258); Activity 1 (258); Activity 2 (258) **12:** Activity 1 (274); Activity 3 (274) **13:** Activity 1 (297); Activity 2 (297); Activity 3 (297) **14:** Activity 1 (316); Activity 2 (316); Activity 3 (316) **15:** Social Studies Connection (343); Activity 1 (343); Activity 3 (343) **16:** Tests and Assessments (347); Assessing the Solution (360); Social Studies Connection (361); Activity 1 (361); Activity 3 (361) **17:** Activity 2 (389) **18:** Activity 2 (405) **Chapter 9 Activity A:** Invention and Innovation (220) **Chapter 10 Activity A:** Design for Space (241) **Chapter 18 Activity A:** Communication Design (407)
	Tech Lab Workbook: **13-1:** Orthographic Sketching (67) **13-2:** Rendering (69) **13-5:** Biorelated Feeder (71) **17-2:** Locker-Organizer Design (91) **Chapter 18 Activity A Worksheet:** Communication Design (95)
Section 4	
Reading	**Text:** **19:** Main Types of Agriculture (414); Agricultural Crops (414); Technology in Agriculture (417); Raising Livestock (426); Special Types of Agriculture (426); Landscape Plants (428); Food Production (430); Artificial Ecosystems (435) **20:** Buildings (445); Civil Structures (448); Constructing a Structure (449); Tower Cranes (457); Social Studies Connection (463) **21:** Conversion and Converters (476); Turbines (480); Energy and Power Transmission (490); Using and Conserving Energy (492) **22:** Communication and Information (502); How We Communicate (503); Communication Technology (503); Types of Communication Systems (507); The Communication Process (520); Information Technologies (531); All Subjects Connection (535); Social Studies Connection (535) **23:** Manufacturing Processes (546); Manufacturing Systems (557); Developing Manufacturing Systems (561); Producing Products (565); Marketing Products (566) *(continued)*

Reading *(continued)*	**Text:** **24:** Goals of Health and Medicine (578) **25:** What Is Transportation? (600); Types of Transportation (601); Flight Simulators (602); Transportation Processes (603); Modes of Transportation (604); Transportation Systems (610)
Writing	**Text:** **19:** Social Studies Connection (438) **21:** Activity 1 (494) **22:** Social Studies Connection (535); Activity 1 (536) **23:** Go Green Manufacturing (574) **25:** Activity 3 (619)
Verbal	**Text:** **22:** Communication and Information (502); How We Communicate (503); The Communication Process (520); Activity 1 (536); Activity 2 (536) **25:** Activity 3 (619)
Math	**Text:** **19:** Mathematics Connections (438) **20:** Mathematics Connection (463) **21:** Mathematics Connections (493) **22:** Mathematics Connection (535) **23:** Mathematics Connection (568) **24:** Mathematics Connection (594) **25:** Mathematics Connection (619)
	Tech Lab Workbook: **20-1:** Estimating Construction Costs (105)
Science	**Text:** **19:** Agriculture and Biotechnology (429); Artificial Ecosystems (435); Science Connections (438) **20:** Science Connections (463) **21:** Energy and Power Transmission (490); Using and Conserving Energy (492); Science Connection (493) **22:** Science Connection (535) **23:** Science Connection (568) **24:** Science Connection (594) **25:** Science Connection (619) **Chapter 19 Activity A:** Agriculture and Biorelated Technology (440) **Chapter 21 Activity A:** Energy-Conversion Technology (495)
	Tech Lab Workbook: **Chapter 21 Activity A Worksheet:** Energy-Conversion Technology (111)

Analytical	**Text:** **19:** Activity 1 (438); Activity 2 (438); Activity 3 (438) **20:** Social Studies Connection (463); Activity 1 (463); Activity 2 (463); Activity 3 (463); Engineering Structure (471) **21:** Social Studies Connection (493); Activity 2 (494); Activity 3 (494); Activity 4 (494) **22:** All Subjects Connection (535); Activity 2 (536); Activity 3 (536); Graphic Design (541) **23:** Social Studies Connections (568); Activity 1 (568); Activity 2 (568); Activity 3 (568) **24:** Social Studies Connection (594); Activity 1 (594); Activity 2 (594); Activity 3 (594) **25:** Social Studies Connection (619); Activity 1 (619); Activity 2 (619) **Chapter 20 Activity A:** Construction Technology (465) **Chapter 22 Activity A:** Information and Communication Technology (538) **Chapter 23 Activity A:** Manufacturing Technology (570) **Chapter 24 Activity A:** Medical Technology (596) **Chapter 25 Activity A:** Transportation Technology (621)
	Tech Lab Workbook: **20-1:** Estimating Construction Costs(105) **20-2:** Model-Bridge Design Contest (107) **Chapter 22 Activity A Worksheet:** Information and Communication Technology (117) **22-1:** Billboard Message (123) **23-1:** Recipe Holder (131) **Chapter 25 Activity A Worksheet:** Transportation Technology (139) **25-1:** Hovercraft Design (141)
Section 5	
Reading	**Text:** **26:** The World of Change (630); Earthquake Protection (632); Technology and Change (633); Negative Impacts (639); Activity 2 (650); Activity 3 (650) **27:** Futurists (658); Forecasting (658); The Future of Technology (663)
Writing	**Text:** **26:** Environmental Focus (654) **27:** Language Arts Connection (668); Social Studies Connection (668); Activity 2 (668) **Chapter 26 Activity A:** Technological Impacts (652)
	Tech Lab Workbook: **27-1:** Future Telling (149)
Verbal	**Text:** **26:** Activity 4 (650)
Math	**Text:** **26:** Mathematics Connection (650)
Science	**Text:** **26:** Science Connection (650)

Analytical	**Text:** **26:** Social Studies Connection (650); Activity 1 (650); Activity 2 (650); Activity 3 (650); Environmental Focus (654) **27:** Activity 1 (668); Activity 3 (668) **Chapter 26 Activity A:** Technological Impacts (652)
	Tech Lab Workbook: **27-1:** Future Telling (149) **27-2:** Products of the Future (151)

Scope and Sequence Chart

In planning your program, you might want to use this Scope and Sequence Chart to identify the major concepts presented in each chapter of the *Exploring Design, Technology, & Engineering* text. Refer to the chart to select topics meeting your curriculum needs. The chart is divided into five sections. Within these sections, bold numbers indicate the chapters in which the concepts are found. Topics and their corresponding page numbers follow the chapter numbers for easy reference to the text.

Section 1
Core Concepts of Technology **1:** Technology and Knowledge (28); Technology and Action (30) **2:** System Goals (40); Technology and Knowledge (43); Technology and Time (44)
Connections to Other Fields of Study **1:** Technology and History (26); Technology and Language (31) **3:** The Stone Age—Up to about 3500 BC (60); The Bronze Age—About 3500 to 1000 BC (61); The Iron Age—1000 to 500 BC (63); The Industrial Age—1750 to the Late 1900s (64); The Information Age—Late 1900s to Present Day (67)
Technology and Its Impacts **1:** Assessing Actions (31) **2:** Technology and People (42) **3:** Historical Context (59); Societal Context (68); Scope of Society (68); Technology and the Individual (69); Technology and Society (70)
Design and Problem Solving in Technology **1:** Designing Actions (30)
Technological Tools and Materials **2:** Technology and Natural Resources (42); Technology and Machines (43)
Technological Systems and Processes **1:** Producing Actions (31) **2:** System Inputs (41); System Processes (44); Production Processes (45); System Outputs (48); System Feedback (50)
Applications of Technology **1:** Using Actions (31) **2:** Technology and Energy (44) **3:** Functional Context (72)

Technological Enterprises **2:** Technology and Financial Resources (44); Management Processes (48)
Section 2
Core Concepts of Technology **6:** Energy and Work (139); Energy and Power (140); The Nature of Energy (142); Forms of Energy (142); Primary and Secondary Energy Sources (143); Common Energy Supplies (144) **7:** Data, Information, and Knowledge (168); Scientific Knowledge (170); Technological Knowledge (171); Gaining Information and Knowledge (172) **8:** Time and Technology (196)
Connections to Other Fields of Study **7:** Scientific Knowledge (170); The Information Age (172)
Technology and Its Impacts **6:** Alternate Energy Sources (151) **8:** Work and a Changing Society (184); Jobs and Technology (185); Jobs in the Information Age (188); Jobs in Technology (189); People and Jobs (191)
Technological Tools and Materials **4:** Toolmakers (80); Tools (81); Measuring Tools (82); Cutting Tools (83); Drilling Tools (86); Gripping Tools (86); Pounding Tools (88); Polishing Tools (89); Mechanisms (89); Levers (90); Wheel and Axles (92); Pulleys (92); Inclined Planes (93); Wedges (93); Screws (93); Machines (94); Turning Machines (95); Drilling Machines (96); Milling and Sawing Machines (96); Planing and Shaping Machines (97); Grinding and Sanding Machines (97); Shearing Machines (97) **5:** Types of Materials (111); Engineering Materials (112); Metals (112); Ceramics (113); Polymers (113); Composites (115); Obtaining Material Resources (115); Renewable Resources (115); Exhaustible Resources (119); Processing Raw Materials (120); Properties of Materials (123); Physical Properties (124); Mechanical Properties (124); Chemical Properties (124); Thermal Properties (124); Electrical and Magnetic Properties (125); Optical Properties (126); Acoustical Properties (126) **6:** Fossil Fuels (145); Waterpower (148); Nuclear Energy (149); Solar Energy (151); Wind Energy (152); Ocean Energy (152); Geothermal Energy (153); Biomass Energy (153)
Technological Systems and Processes **5:** Processing Raw Materials (120); Mechanical Processing (121); Thermal Processing (122); Chemical Processing (122); Products of Processing (123)
Applications of Technology **6:** Waterpower (148); Nuclear Energy (149); Solar Energy (151); Wind Energy (152); Ocean Energy (152); Geothermal Energy (153); Biomass Energy (153)
Technological Enterprises **8:** Jobs and Technology (185); Jobs in the Information Age (188); Jobs in Technology (189); People and Jobs (191); Goals, Interests, and Traits (191); Getting Career Information (193); Advancing in a Career (195); Money and Technology (198)

Section 3

Core Concepts of Technology

10: What Is Design? (225)

11: Needs and Wants (247)

12: Types of Research (262)

15: What Is a Model? (320); Why Are Models Used? (320)

Connections to Other Fields of Study

15: Mathematical Models (323)

Design and Problem Solving in Technology

9: Invention (206); Discovery vs Inventions (208); Categories of Inventions (208); The Invention Process (210); Identifying the Problem (211); Collecting Information (211); Thinking Creatively (211); Experimenting (211); Reviewing (212); Inventors (212); Reasons for Inventing (213); Invention Teams (214); Innovations (216)

10: What Is Design? (225); Types of Design (226); Artistic Design (226); Engineering Design (226); Steps of the Engineering Design Process (228); Identifying the Problem (230); Researching the Problem (231); Creating Solutions (233); Selecting and Refining the Solution (233); Modeling the Solution (235); Testing the Solution (236); Communicating the Solution (236); Improving the Solutions (237)

11: Problems and Opportunities (248); Types of Problems (249); Recognizing Problems (250); Problem Statements (252); Criteria (252); Constraints (254); Design Briefs (256)

13: Ideation (277); Brainstorming (280); Graphic Organizers (280); Questioning (282); Sketching (283); Rough Sketches (283); Refined Sketches (284); Refining Techniques (292); Types of Sketches (293)

14: Selecting Solutions (302); Aspects of Design (303); Appearance (303); Function (303); Design Sheets (308); Evaluation Grids (308); Refining Solutions (310); Elements of Design (310); Principles of Design (312); Renderings (314); Detailed Sketches (314)

15: Communication (320); Experimentation (321); Organization (322); Types of Models (322); Graphic Models (322); Computer Models (324); Physical Models (326)

16: Using the Criteria (348); Testing for Certain Characteristics (349); Function (349); Human Engineering (351); Appearance (354); Safety (354); Testing Environments (355)

17: Drawings (370); Types of Working Drawings (374); Detailed Drawings (374); Assembly Drawings (374); Schematic Drawings (374); Drawing Classifications (376); Orthographic Drawings (376); Pictorial Drawings (379)

18: Reasons for Improvements (396); Function (399); Safety (399); Sources of Improvement (400)

Technological Tools and Materials

15: Modeling Materials (329); Sheet Materials (330); Found Objects (335); Sculpting and Molding Materials (336); Adhesives (337); Modeling Tools (339)

16: Collect Materials and Equipment (359)

17: Communication Documents (370); Bills of Materials (370); Drawings (370); Specification Sheets (370); Mechanical Drawing Tools (382); Traditional Tools (382); Computer-Aided Design (CAD) Tools (384)

Technological Systems and Processes
9: The Invention Process (210) **10:** Steps of the Engineering Design Process (228) **12:** Library Research (262); Survey Research (264); Experimental Research (265); The Research Process (267); List the Problem (267); Gather Data (267); Analyze Data (271); Make a Conclusion (272) **13:** The Sketching Process (284) **14:** External Selection (302); Internal Selection (302); Production (307) **15:** Making a Model (340) **16:** Tests and Assessments (347); The Testing Procedure (358); Determine the Purpose (358); Choose the Test (358); Conduct the Test and Gather Data (359); Analyze the Data (360); Evaluate the Results (360); Assessing the Solution (360) **17:** Making Mechanical Drawings (386); Documenting and Presenting the Solution (387) **18:** Making Improvements (404)
Applications of Technology
9: Social Inventions (213); Financial Inventions (214); Leisure Inventions (214)
Technological Enterprises
9: Financial Inventions (214) **14:** Finances (307) **16:** Economics (352) **18:** Profit (397); Competition (398); Consumers (400); Competitors (401); Government Organizations (402); Internal Sources (404)
Section 4
Core Concepts of Technology
25: What Is Transportation? (600)
Connections to Other Fields of Study
19: Agriculture and Biotechnology (429)
Technology and Its Impacts
21: Using and Conserving Energy (492)
Design and Problem Solving in Technology
23: Designing Tooling (563)
Technological Tools and Materials
19: Power, or Pulling, Equipment (417); Tillage Equipment (418); Planting Equipment (420); Pest-Control Equipment (420); Irrigation Equipment (421); Harvesting Equipment (423); Storage Equipment (425) **22:** Communication and Information (502); Computer Hardware (532); Computer Software (533) **23:** Selecting Equipment (562); Arranging Equipment (562); Designing Tooling (563)

Technological Systems and Processes

20: Constructing a Structure (449); Preparing to Build (450); Preparing the Site (453); Setting Foundations (453); Building the Framework (454); Enclosing the Structure (458); Finishing the Interior (459); Finishing the Site (461)

21: Conversion and Converters (476); Energy and Power Transmission (490); Mechanical Transmission (490); Fluid Transmission (491); Electrical Transmission (491); Using and Conserving Energy (492)

22: How We Communicate (503); Types of Communication Systems (507); The Communication Process (520); Encoding (520); Transmitting (521); Receiving (522); Decoding (522); Storage (523); Retrieval (524); Producing Published Messages (526); Producing Film Messages (529)

23: Primary Processes (547); Secondary Processes (549); Manufacturing Systems (557); Custom Manufacturing (557); Intermittent Manufacturing (557); Continuous Manufacturing (558); Flexible Manufacturing (559); Developing Manufacturing Systems (561); Selecting Operations (561); Sequencing Operations (561); Controlling Quality (563); Testing the System (564); Producing Products (565)

25: Transportation Processes (603); Transportation Systems (610); Vehicle Systems (610); Support Systems (616)

Applications of Technology

19: Main Types of Agriculture (414); Agricultural Crops (414); Technology in Agriculture (417); Raising Livestock (426); Special Types of Agriculture (426); Food Production (430); Food Processing (431); Food Preservation (432); Artificial Ecosystems (435)

20: Buildings (445); Residential Buildings (445) Commercial Buildings (446); Industrial Structures (447); Public Buildings (447); Religious Buildings (447); Civil Structures (448)

21: Conversion and Converters (476); Mechanical Converters (478); Thermal Converters (479); Chemical Converters (484); Solar Converters (485); Fluid Converters (488)

22: Communication Technology (503); Graphic Communication (504); Wave (Electronic) Communication (504); Graphic Communication versus Electronic Communication (506); People-to-People Communication (508); People-to-Machine Communication (518); Machine-to-People Communication (519); Machine-to-Machine Communication (519); Information Technologies (531)

23: Manufacturing Processes (546)

24: Goals of Health and Medicine (578); Prevention Programs (578); Dealing with Illness and Injury (584); Using Technology in Medicine (584)

25: Types of Transportation (601); Modes of Transportation (604); Land Transportation (604); Water Transportation (607); Air Transportation (608); Space Transportation (609); Intermodal Transportation (610)

Technological Enterprises

23: Marketing Products (566)

Section 5

Technology and Its Impacts

26: The World of Change (630); Technology and Change (633); Negative Impacts (639); Impacts on Resources (640); Impacts on the Environment (642); Impacts on People (644)

27: Futurists (658); The Future of Technology (663)

Technological Tools and Materials
26: Materials and Change (633); Tools and Change (634); Impacts on Resources (640) **27:** Energy and the Future (664)
Technological Systems and Processes
27: Forecasting (664); Trending (660); Surveying (660); Creating Scenarios (661); Modeling (662)
Applications of Technology
26: Communication and Change (634); Transportation and Change (637); Construction and Change (638) **27:** Agriculture and the Future (663); Construction and the Future (664); Communication and the Future (665); Medicine and the Future (666); Manufacturing and the Future (666); Transportation and the Future (666)
Technological Enterprises
26: Production and Change (639)

Section 1
Scope of Technology

Focus: Introduction to technology, technological systems, and contexts in which technology is applied.

- ❑ Chapter 1: What Is Technology?
- ❑ Chapter 2: Technology as a System
- ❑ Chapter 3: Contexts of Technology

Teaching Materials

Text

Chapter 1, pages 24–35
- Chapter Activities, page 33
- Test Your Knowledge, page 33
- Chapter 1 Activity A—What Is Technology?, page 35

Chapter 2, pages 36–57
- Chapter Activities, page 52
- Test Your Knowledge, page 52
- Chapter 2 Activity A—Using Technology to Make a Product, pages 53–57

Chapter 3, pages 58–75
- Chapter Activities, page 74
- Test Your Knowledge, page 75

Tech Lab Workbook

Chapter 1
- Chapter 1 Study Questions, pages 7–8

Chapter 2
- Chapter 2 Activity A Worksheet—Using Technology to Make a Product, pages 9–12
- Student Activity 2-1—Marble Counter, page 13
- Student Activity 2-2—Technological Systems, pages 15–16
- Chapter 2 Study Questions, pages 17–18

Chapter 3
- Chapter 3 Study Questions, pages 19–20

Teacher's Resources

Chapter 1
- Alternative Activity—Diamond Kite, pages 49–54
- Lesson Slide 1-1—Technology Is Knowledge (CD only)
- Lesson Slide 1-2—Technological Actions (CD only)
- Lesson Slide 1-3—Types of Knowledge (CD only)

Chapter 2
- Reproducible Master 2-1—The Universal Systems Model, page 55
- Reproducible Master 2-2—Types of Results for Technological Systems, page 56
- Reproducible Master 2-3—Systems, page 57
- Lesson Slide 2-1—Technological Systems (CD only)
- Lesson Slide 2-2—Inputs to Technological Systems (CD only)
- Lesson Slide 2-3—Processes of Technological Systems (CD only)
- Lesson Slide 2-4—Outputs of Technological Systems (CD only)
- Lesson Slide 2-5—Feedback in Technological Systems (CD only)

Chapter 3
- Lesson Slide 3-1—Functional Contexts (CD only)

Introduction

This section sets the stage for the entire study of technology. Section 1 first introduces the class to the term *technology*. This section then explains the difference between technology and science. "Scope of Technology" presents technology as a system to extend human potential. This section describes designing, producing, using, and assessing as the major actions of technology. Section 1 discusses technological systems in terms of inputs, processes, outputs, feedback, and societal influences. Finally, it introduces agriculture, construction, energy conversion, information and communication, manufacturing, medicine, and transportation as major contexts in which technology is applied.

Objectives of the Section

At the end of Section 1, each student should be able to do the following:

- ❑ Understand the term *technology*.
- ❑ Compare technology and science.
- ❑ Explain technology as a system.
- ❑ Recall the parts of a technological system and their functions.
- ❑ Give examples of production and management processes.
- ❑ Summarize the goals of technology.
- ❑ Summarize the characteristics of the major contexts in which technology is applied.

Suggested Schedule and Length of the Section (5–10 Days)

Section 1 should introduce students to technology. Have them complete the following assignments:

1. Read Chapters 1–3 in their textbooks and answer associated study questions.
2. Participate in presentations and discussions on technology, technological systems, and technological contexts.
3. Complete one or two laboratory activities and follow-up sheets (in the Tech Lab Workbook).

Schedule for Section 1—9-Week and 12-Week Classes	
Day 1	School start-up activities Textbook and Tech Lab Workbook distribution Discussion: "What will we study in this class?" Assign Chapter 1: "What Is Technology?" to be read outside of class.
Day 2	Discussion: "What is technology?" Assign Chapter 2: "Technology as a System" to be read outside of class. Introduce activity.
Day 3	Work on activity. Assign Chapter 3: "Contexts of Technology" to be read outside of class.
Day 4	Discussion: "Technological systems and contexts." Work on activity.
Day 5	Complete and summarize activity and section.

Schedule for Section 1—18-Week and 36-Week Classes	
Day 1	School start-up activities Textbook and Tech Lab Workbook distribution Discussion: "What will we study in this class?"
Day 2	Discussion: "What is technology?" Assign Chapter 1: "What Is Technology?" to be read outside of class. Introduce first activity.
Day 3	Work on activity.
Day 4	Work on activity. Assign Chapter 2: "Technology as a System" to be read outside of class.
Day 5	Discussion: "What is a system?" Summarize and discuss activity.
Day 6	Introduce second activity. Assign Chapter 3: "Contexts of Technology" to be read outside of class.
Day 7	Work on activity.
Day 8	Discussion: "What are technological contexts?" Work on activity.
Day 9	Complete and summarize activity and section.

Preparing to Teach the Section

You have several important tasks to complete before this section can be presented to students.

1. Carefully read this Teacher's Resource to develop an understanding of the scope and sequence of the course.
2. Read the textbook cover to cover to develop an understanding of the content.
3. Prepare presentation and discussion plans. Make notes for four 10- to 15-minute sessions on the following topics:
   A. What is technology? (Textbook, Chapter 1)
   B. Technology as a system. (Textbook, Chapter 2)
   C. What are technological contexts? (Textbook, Chapter 3)
   D. Technology—A summary.
4. Select one or two appropriate laboratory activities from the following options:
   A. What Is Technology? (Chapter 1), Textbook, page 35.
   B. Diamond Kite (Chapter 1), Teacher's Resource, pages 49–54.
   C. Using Technology to Make a Product (Chapter 2), Textbook, pages 53–57.
   D. Marble Counter (Chapter 2), Tech Lab Workbook, page 13.
   E. Technological Systems (Chapter 2), Tech Lab Workbook, page 15.
5. Make necessary material and supply preparations for presenting the laboratory activity or activities.
6. Work through the laboratory activity or activities and student study questions presented in the textbook, Tech Lab Workbook, and Teacher's Resource.

Presenting the Section

All sections in this program follow a basic format. They include the following elements:

- ❑ Discussions and presentations to introduce and reinforce basic information. Plan to limit these sessions to 10- or 15-minute periods distributed across several days.
- ❑ Student study activities included in the Key Words and Test Your Knowledge sections for each textbook chapter and study questions in the Tech Lab Workbook.
- ❑ Laboratory activities. These activities shift the basic information from the abstract to the concrete. Refer to the suggested activities contained in the Activities sections of the textbook, activities in the Tech Lab Workbook, and alternative activities in the Teacher's Resource. Remember to develop and demonstrate safe procedures for completing each activity.
- ❑ Laboratory summary sheets to ensure that the students can relate the activities to the course content. These sheets are contained in the Tech Lab Workbook.

Adjustments for Students with Special Needs and Gifted Students

The five activities suggested for this section lend themselves to group work. One way to adjust instruction for differing abilities and varying interests is to organize the groups with a mix of students in each one. Each group can include gifted students, average students, and students with special needs. Consider grouping those who think logically and sequentially with random, intuitive thinkers. Some students might have highly developed artistic abilities. Think about pairing them with others who have little or no artistic talent. Give special attention and individual help to those students who require more structure. If the activity is highly structured for all students, however, the gifted students might lack the challenge they need to do their best work.

Evaluation Procedures

Technology education stresses the process and creative answers. Quality (craftsmanship) of the final product is of secondary importance. Therefore, students should be evaluated on the following criteria:

- ❑ Contributions to group activity. This evaluation is possible with daily logs (diaries)

the students maintain, teacher observations, and student feedback.
- ❑ Laboratory sheets, sketches, and designs completed during the activity.
- ❑ Participation in discussions conducted during the laboratory activity.
- ❑ Scores on quizzes, chapter questions, or other activities.

Implementing the Laboratory Activities

Five different activities have been developed for this section. These are simply suggested activities. Any activity introducing technology can be substituted. The time constraints for this section restrict the program to one or two relatively short activities.

Procedures for Directing Each Suggested Activity

Chapter 1: What Is Technology?, Textbook, page 35

This activity is designed to allow the students to use their creativity to design an energy-conversion system. They will use the mechanical energy from wind to lift a weight. There is no right answer. To use this activity, divide the class into groups of two to four students. Have each group complete the following steps:

1. Define the problem they are attempting to solve.
2. List any restrictions they face, such as materials that can be used (a list is included with the activity description in the textbook) and the strength of the wind.
3. Develop sketches of two to five ways to solve the problem.
4. Select and refine the best solution.
5. Build a prototype of the solution using only the materials listed in the textbook. (Limiting the materials introduces a common restriction all product designers face.)
6. Demonstrate the solution.
7. List ways to improve the design.

Chapter 1: Diamond Kite, Teacher's Resource, pages 49–54

This activity is included in this Teacher's Resource as a substitute for the clay-tile activity in the textbook or a second activity for 18- and 36-week classes. If this activity is used, the students will work in small groups, using only pencils, rulers, a backsaw, and scissors, to produce a kite with a low-tech production system. They will then use templates and jigs to make another kite, using a high-tech system. You need to make the following jigs and fixtures for each group to use:

- ❑ A template to mark out the cover.
- ❑ An alignment jig that positions the spine and spar so they can be assembled without measuring.
- ❑ A jig to hold the spine and spar so the notches can be cut quickly.

Also, the laboratory sheets for the clay tile need to be modified to ask questions about the kite and its quality.

Chapter 2: Using Technology to Make a Product, Textbook, pages 53–57, and Tech Lab Workbook, pages 9–12

This activity allows students to explore technology by, first, producing a product without tools and, then, making the same product using technological aids. "Using Technology to Make a Product" is designed to be prescriptive in nature, with the procedure provided for the students. During Task 1 of the activity, the students form and decorate 1/4″ × 4″ × 4″ clay tiles using their hands. Have students complete the following steps:

1. Read the directions for Task 1 of Chapter 2 Activity A on page 53 of the textbook.
2. Form as many tiles as they can in a set time period, such as 10 minutes.
3. Place their tiles with the other ones the class members produce.
4. Complete the first section of the Chapter 2 Activity A Worksheet, found on page 9 of the Tech Lab Workbook.

5. Discuss the problems with producing a product without technology.
6. Recommend ways to develop technology (tools) to increase the speed of producing the tiles.

During Task 2, divide the class into groups of five students. Demonstrate the process the students will use to make tiles using technology. Have each student record the steps on the activity procedure sheet found on page 10 of the Tech Lab Workbook. Each group should complete the following steps:

1. Read the directions for Task 2 of Chapter 2 Activity A on page 55 of the textbook.
2. Collect the supplies:
   - ❑ Two 1/4″ × 3/4″ × 12″ wood strips.
   - ❑ A rolling pin.
   - ❑ A tile cutter.
   - ❑ A tile-decorating press.
   - ❑ Potter's clay.
   - ❑ A sponge.
   - ❑ A can of water.
3. Create groups of students by counting off numbers from one to five.
4. Produce 12 tiles using the procedure.
5. Analyze the tiles.
6. Complete the Chapter 2 Activity A Worksheet, found on page 11 of the Tech Lab Workbook.
7. Discuss the problems with producing the products using technology.
8. Complete the Activity Summary, found on page 11 of the Tech Lab Workbook.
9. Recommend ways to improve the technology (tools) used to produce the product.

Chapter 2: Marble Counter, Tech Lab Workbook, page 13

This activity is designed to be an open-ended problem-solving activity. The students should be provided the design brief and encouraged to develop a unique solution to the problem. The activity can be done individually. "Marble Counter" will work better, however, with groups of two to four students. Have the students complete the following steps:

1. Define the problem they are attempting to solve.
2. List any restrictions they face, such as marble size and cost of the solution.
3. Develop sketches of two to five ways to solve the problem.
4. Select and refine the best solution.
5. If time permits, build a prototype of the solution.

Chapter 2: Technological Systems, Tech Lab Workbook, pages 15–16

"Technological Systems" is designed to allow the students to explore the components of technological systems. This activity can be done individually or in small groups. Have the students complete the following steps:

1. Read the directions for Student Activity 2-2 on page 15 of the Tech Lab Workbook.
2. Complete the Student Activity 2-2 worksheet.
3. Discuss their answers with the rest of the class.

Answer Key for Section 1

Chapter 1—What Is Technology?

Test Your Knowledge, Textbook, page 33

1. Answers will vary. Some acceptable answers are pollution, crime, unemployment, and global warming.
2. Technological knowledge used in the past can be applied to current situations to help people lead better and easier lives.
3. B. Knowledge about developing and making products.
4. A. Scientific knowledge.
   B. Humanities knowledge.
   C. Descriptive knowledge.
   D. Technological knowledge.
5. A. Knowledge about using machines and systems to grow, harvest, and preserve foods and to raise and use livestock.
   B. Knowledge about using machines and systems to erect buildings and other structures on the sites where they will be used.
   C. Knowledge about using machines and systems to convert, transmit, and apply energy.

D. Knowledge about using machines and systems to collect, process, and exchange information and ideas.
E. Knowledge about using machines and systems to treat diseases and maintain the health of living beings.
F. Knowledge about using machines and systems to convert natural materials into products.
G. Knowledge about using machines and systems to move people and cargo.

6. designing actions
7. producing actions
8. using actions
9. assessing actions
10. vocabulary or words, technical drawings

Study Questions, Tech Lab Workbook, pages 7–8

1. Answers will vary. Some acceptable answers are a computer, bicycle, telephone, microwave, and television.
2. Answers will vary. Some acceptable answers are the following:
   Positive: quick communication, communication from any location, ease of use.
   Negative: distraction from other tasks, such as driving; cost; lack of privacy when using.
3. Answers will vary. Some acceptable answers are the following:
   Obsolete skills: harness making, wheel repair by a wheelwright, wagon repair.
   New skills: steering the vehicle, engine repair, vehicle servicing.
4. A. Scientific
   B. Technological
5. Agricultural, construction, energy, information and communication, manufacturing, medical, and transportation.
6. B. Designing.
7. C. Producing.
8. D. Using.
9. A. Assessing.
10. vocabulary, technical drawings

Chapter 2—Technology as a System

Test Your Knowledge, Textbook, page 52

1. C. A group of parts working together to complete a task.
2. Natural systems appear in nature without human interference. Technological systems are human made.
3. Goals, inputs, processes, outputs, and feedback. Goals are the reasons for developing and operating the system. Inputs are the resources the system uses. Processes are the actions taken. Outputs are the results obtained. Feedback includes the adjustments made.
4. To meet human needs and wants, make a profit, and positively contribute to society.
5. People, natural resources or materials, tools or machines, knowledge or information, energy, financial resources or money, and time.
6. Production and management.
7. managers
8. intended
9. undesirable
10. A closed-loop system is a built-in part of the device. External control is needed in an open-loop system.

Study Questions, Tech Lab Workbook, pages 17–18

1. system
2. A. T
   B. T
   C. N
   D. T
   E. N
   F. N
   G. T
3. E. Goal.
4. A. Input.
5. B. Process.
6. C. Output.
7. D. Feedback.
8. Meeting the needs and wants of people.

9. People, natural resources or materials, tools or machines, knowledge or information, energy, financial resources or money, and time.
10. Production
11. Management
12. undesirable
13. closed, open

Chapter 3—Contexts of Technology

Test Your Knowledge, Textbook, page 75

1. Historical, societal, and functional.
2. During this stage of human history, stone was used as the main material for tools.
3. Bronze
4. D. The period of history when building techniques improved, road systems were developed, and new cooking ware and agricultural tools were created.
5. Industrial
6. Information
7. global
8. Answers will vary. An acceptable answer might state that technology has led people to expect systems and devices that are immediate, constant, flexible, inexpensive, and easy-to-use.
9. True.
10. A. Developing and using devices and systems to plant, grow, and harvest crops and raise livestock.
    B. Developing and using devices and systems to gather, process, and share information and ideas.
    C. Using systems and processes to erect structures on the sites where they will be used.
    D. Developing and using devices and systems to convert, transmit, and use energy.
    E. Developing and using systems and processes to convert materials into products.
    F. Developing and using devices and systems promoting health and curing illnesses.
    G. Developing and using devices and systems to move people and cargo.

Study Questions, Tech Lab Workbook, page 19–20

1. A. Historical—Viewing technology as it has developed over time.
   B. Societal—Viewing how society's goals and values have impacted the development of technology.
   C. Functional—Viewing technology in terms of the major function it performs in human endeavor.
2. A. Stone Age—Crude tools made from stones and sticks.
   B. Bronze Age—Simple metal tools produced through the smelting of copper.
   C. Iron Age—Tools made from iron.
   D. Industrial Age—Water-, steam-, and electric-powered machines; mass production.
   E. Information Age—Computer–data processing machines.
3. Answers will vary, depending on the technology the student chooses.
4. A. Agricultural technology.
   B. Communication and information technology.
   C. Construction technology.
   D. Energy and power technology.
   E. Manufacturing technology.
   F. Medical technology.
   G. Transportation technology.

Captions for Reproducible Masters

- ❑ **Reproducible Master 2-1.** The Universal Systems Model—Systems consist of inputs, processes, outputs, and feedback.
- ❑ **Reproducible Master 2-2.** Types of Results for Technological Systems—Systems can have positive and negative outcomes.
- ❑ **Reproducible Master 2-3.** Systems—A worksheet to help students learn the basic parts of a system.

Captions for Lesson Slides (CD Only)

- ❑ **Lesson Slide 1-1.** Technology Is Knowledge—Technology is human know-ledge about using materials to modify the natural and human-made environments.
- ❑ **Lesson Slide 1-2.** Technological Actions—Technology involves designing, producing, using, and assessing actions.
- ❑ **Lesson Slide 1-3.** Types of Knowledge—The types of knowledge.
- ❑ **Lesson Slide 2-1.** Technological Systems—A system is a group of parts working together to achieve a goal.
- ❑ **Lesson Slide 2-2.** Inputs to Technological Systems—The seven categories of inputs to technological systems are shown.
- ❑ **Lesson Slide 2-3.** Processes of Technological Systems—These processes are used to develop and operate technological systems.
- ❑ **Lesson Slide 2-4.** Outputs of Technological Systems—The possible types of outputs of technological systems are shown.
- ❑ **Lesson Slide 2-5.** Feedback in Technological Systems—An example of a feedback system is shown.
- ❑ **Lesson Slide 3-1.** Functional Contexts—There are many types of technology developed for specific uses. Each type serves a different function. A product, a system, or an environment developed for one situation, however, might be useful to another situation as well.

Name ____________________ Date __________

Period ____________________ Score __________

Chapter 1 Alternative Activity

Diamond Kite

The diamond kite is probably the kite most people first used. This kite is a flat kite that needs a tail to fly properly. A typical diamond kite has five basic parts:

- ❑ **Frame.** This forms the shape of the kite and supports the cover. The frame is made up of two parts:
  - ❑ **Spine.** This is the vertical stick around which the kite is built.
  - ❑ **Spar.** This is the support stick across the spine at a 90° angle.
- ❑ **Cover.** This consists of thin materials, such as paper, plastic, or cloth, covering the frame.
- ❑ **Bridle.** This is one or more strings helping to control the kite. These strings are attached to the spine or spars.
- ❑ **Flying line.** This is the string running from the bridle to the person flying the kite.
- ❑ **Tail.** This is a narrow strip of material, such as paper or plastic, helping to balance the kite in flight.

Materials for a Diamond Kite

- ❑ One sheet of strong paper (36″ × 36″).
- ❑ Two straight wooden sticks or dowels (25″ and 33″).
- ❑ Thin wrapping or garden string.
- ❑ Clear acetate tape or glue.
- ❑ Markers, paint, or crayons.

Making the Diamond Kite with a Low-Tech Production System

Making the Frame

1. Mark the wood strips or dowels for length: one at 25″ and one at 33″.
2. Cut the wood strips or dowels to length.
3. Cut a notch in each end of the strips that is the saw width and 1/4″ deep.

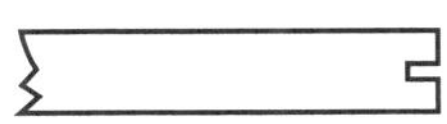

Alternate Activity *(continued)* Name ____________________

4. Place the shorter stick (spar) horizontally across the longer stick (spine) to form a cross. Make sure the horizontal strip is centered so both sides of the crosspiece are equal.

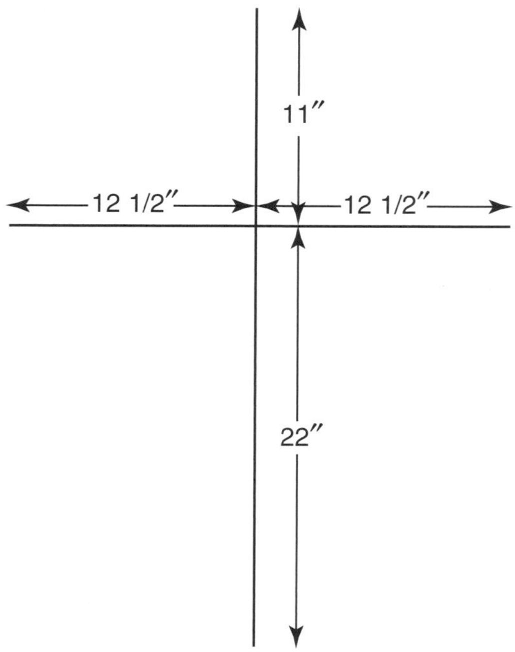

5. Tie the two sticks together with the string. Be sure the sticks are at right angles to each other. To make a stronger joint, apply a drop of glue where the sticks cross.
6. Cut a piece of string long enough to stretch all around the kite frame.
7. Loop the string around the top notch in the frame. Fasten the string by wrapping it around the stick.

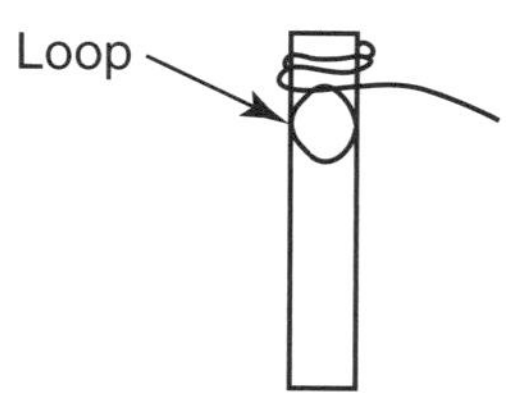

8. Stretch the string through the notch at one end of the crosspiece.
9. Make another loop at the bottom of the frame. Attach it to the frame.
10. Stretch the string through the notch at the other end of the crosspiece and up to the top of the frame.
11. Finish attaching the string by wrapping it a few times around the top of the stick and cutting off what you do not need. *Note:* This string must be tight. The piece of string should not, however, warp the sticks.

Alternate Activity *(continued)* Name ____________________

Making the Cover

1. Lay the paper cover on a flat surface.
2. Lay out the cut and fold line.

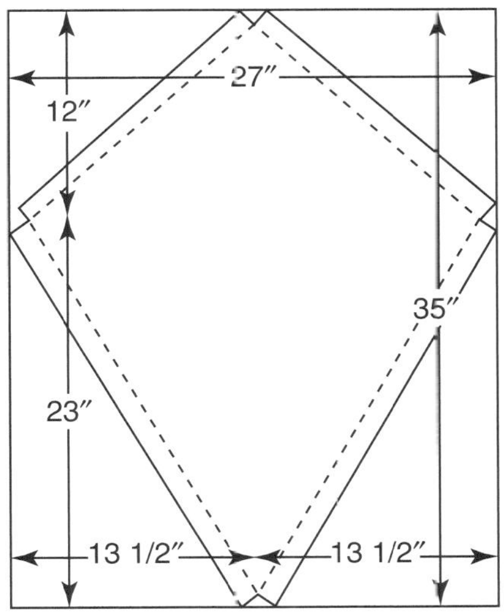

3. Cut out the material.
4. Crease the cover on the fold lines.

Assembling the Kite

1. Fold these edges over the string frame.
2. Glue the paper down so the material is tight.

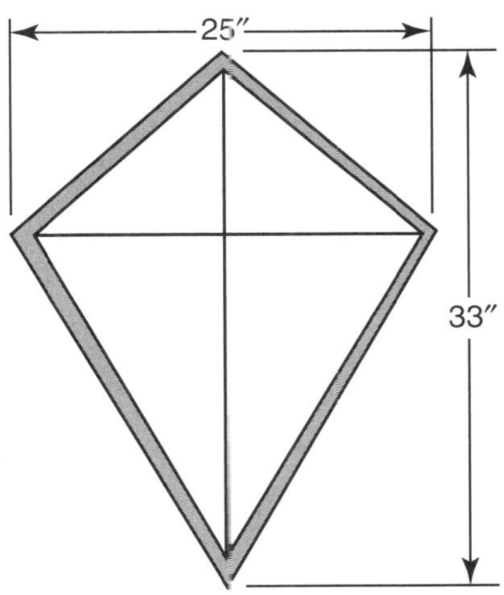

3. Cut a piece of string about 40″ long.
4. Tie one end of the string to the loop at the top of the frame and the other end to the loop at the bottom.
5. Tie a small loop in the string slightly above the intersection of the two crosspieces. This is the kite's bridle. The flying line is attached at the bridle.
6. Make a tail by tying a small strip of fabric about every 4″ along the length of string.
7. Attach the tail to the loop at the bottom of the kite.
8. If you have time, decorate your kite with colored markers, paints, or crayons.

Alternate Activity *(continued)* Name ____________________

Making the Diamond Kite with a High-Tech Production System

Making the Frame

1. Place a strip of wood in the cutting jig. Cut a piece 33″ long.

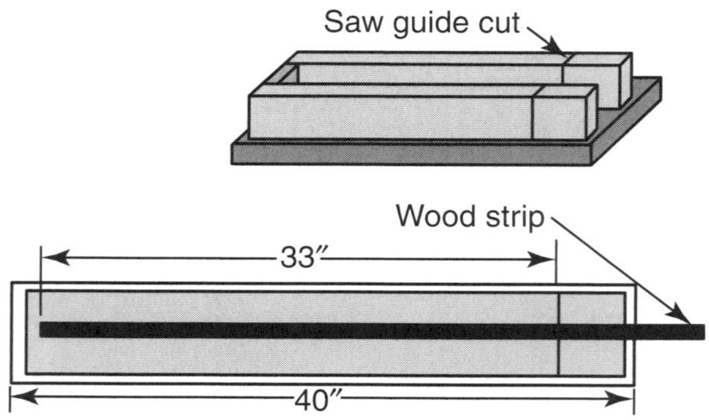

2. Place an 8″ spacer strip in the jig.
3. Place a strip of wood in the cutting jig. Cut a piece 25″ long.

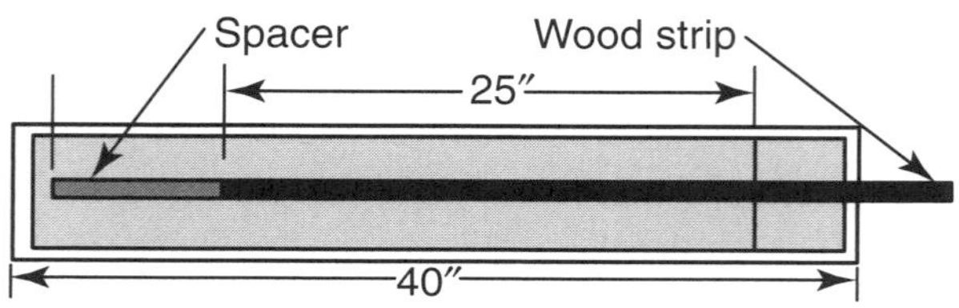

4. Place the ends of the strips in the cutting jig. Cut string notches in all four ends.
5. Place the shorter stick (spar) horizontally in the assembly jig.
6. Place the longer stick (spine) to form a cross.

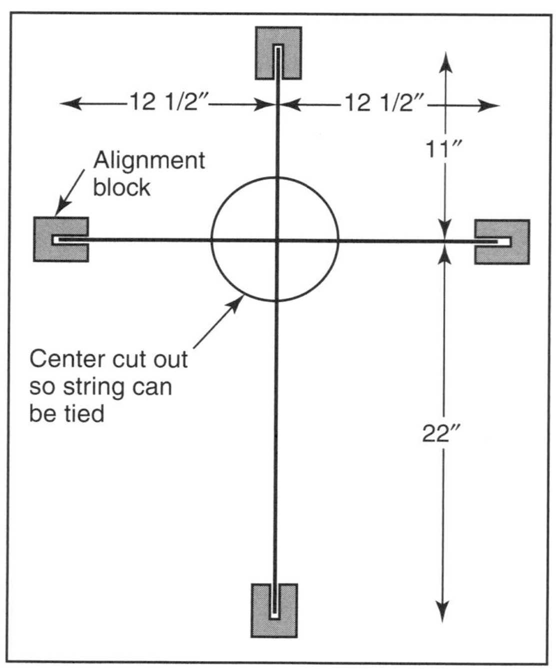

Alternate Activity *(continued)* Name ___________________________

7. Tie the two sticks together with the string. To make a stronger joint, apply a drop of glue where the sticks cross.
8. Cut a piece of string long enough to stretch all around the kite frame.
9. Loop the string around the top notch in the frame. Fasten the string by wrapping it around the stick.

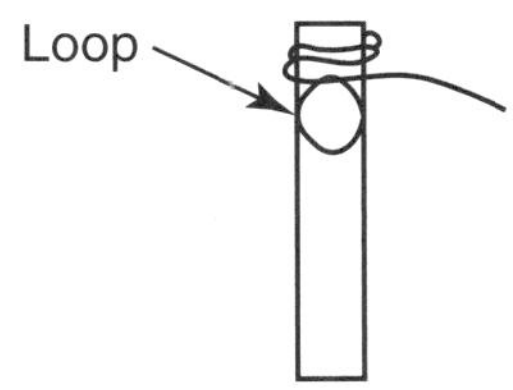

10. Stretch the string through the notch at one end of the crosspiece.
11. Make another loop at the bottom of the frame. Attach it to the frame.
12. Stretch the string through the notch at the other end of the crosspiece and up to the top of the frame.
13. Finish attaching the string by wrapping it a few times around the top of the stick and cutting off what you do not need. *Note:* This string must be tight. The piece of string should not, however, warp the sticks.

Making the Cover

1. Lay the paper cover on a flat surface.
2. Place the cut-marking template over the paper. Draw cutting lines.

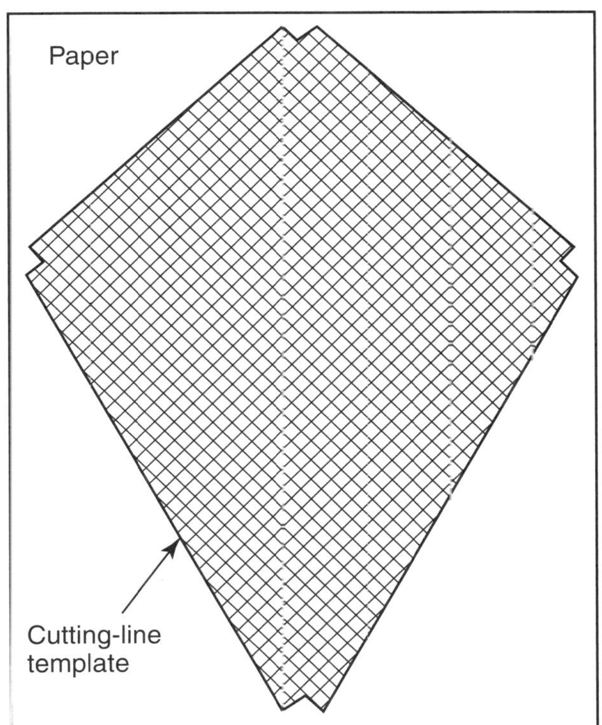

Alternate Activity *(continued)* Name ____________________

3. Place the fold-marking template over the paper. Draw a line where the fold will be.

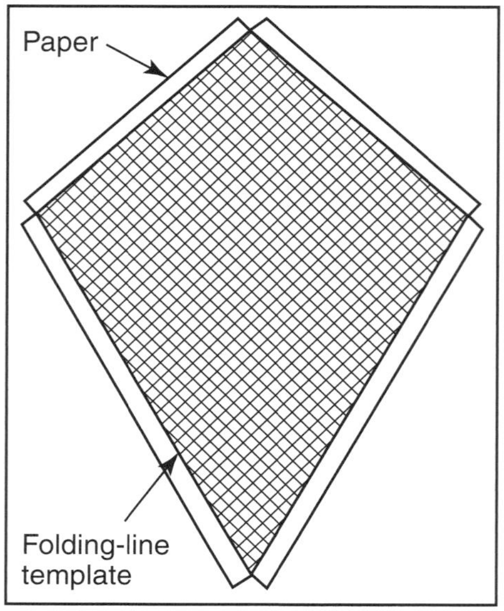

4. Cut out the material.
5. Crease the cover on the fold lines.

Assembling the Kite

1. Fold these edges over the string frame.
2. Glue the paper down so the material is tight.

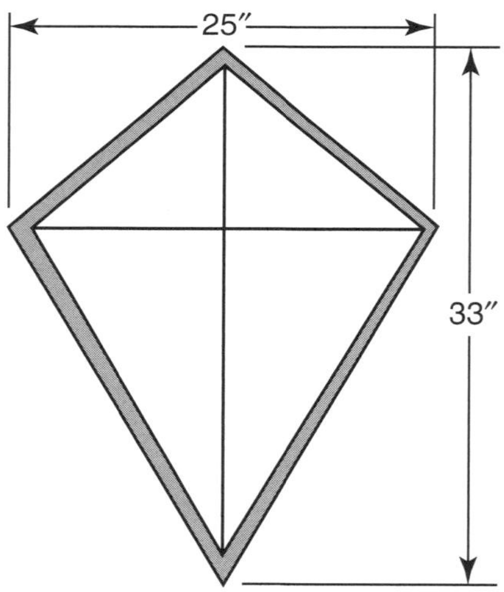

3. Cut a piece of string about 40″ long.
4. Tie one end of the string to the loop at the top of the frame and the other end to the loop at the bottom.
5. Tie a small loop in the string slightly above the intersection of the two crosspieces. This is the kite's bridle. The flying line is attached at the bridle.
6. Make a tail by tying a small strip of fabric about every 4″ along the length of string.
7. Attach the tail to the loop at the bottom of the kite.
8. If you have time, decorate your kite with colored markers, paints, or crayons.

The Universal Systems Model

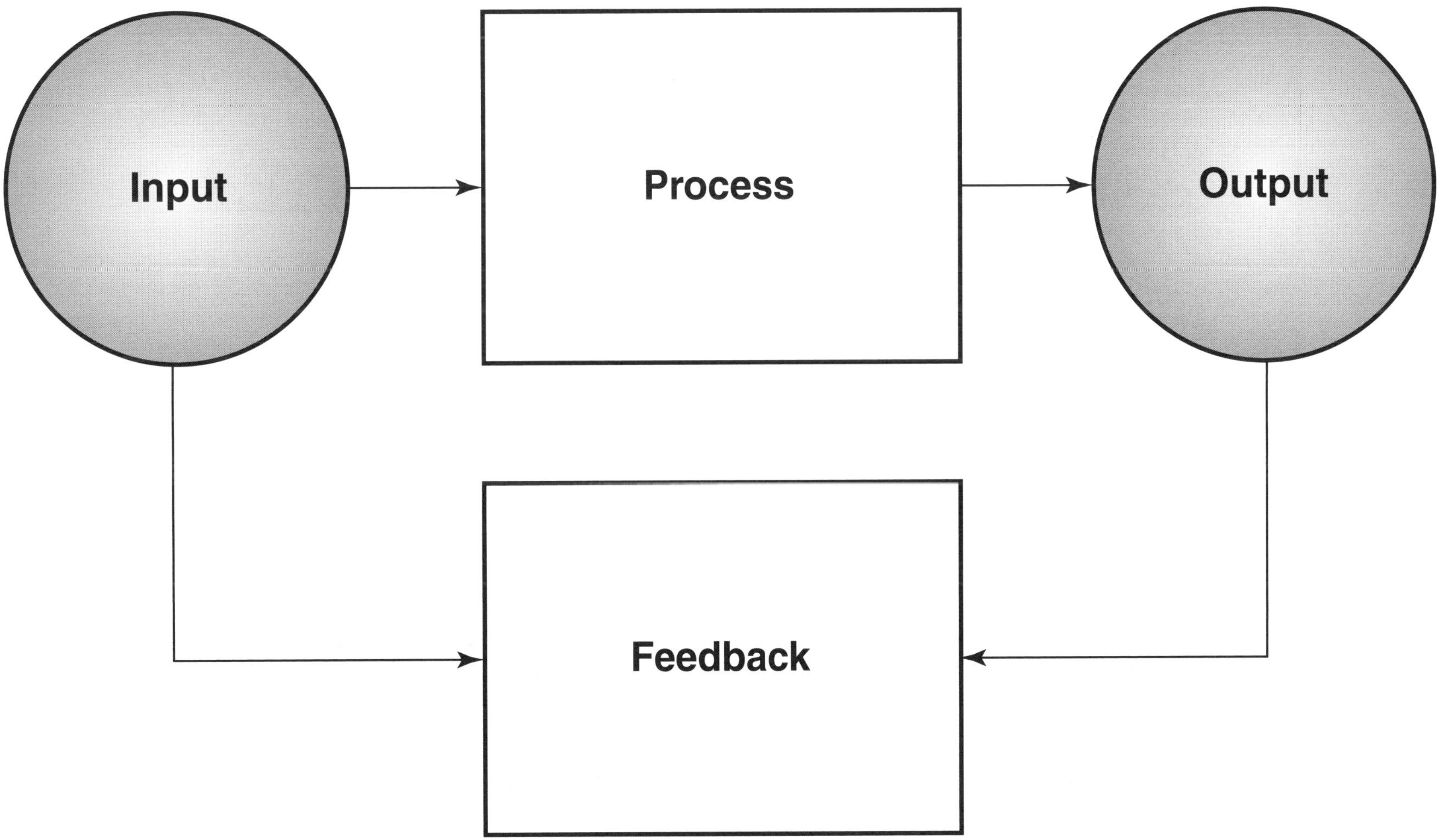

Reproducible Master 2-1

Types of Results for Technological Systems

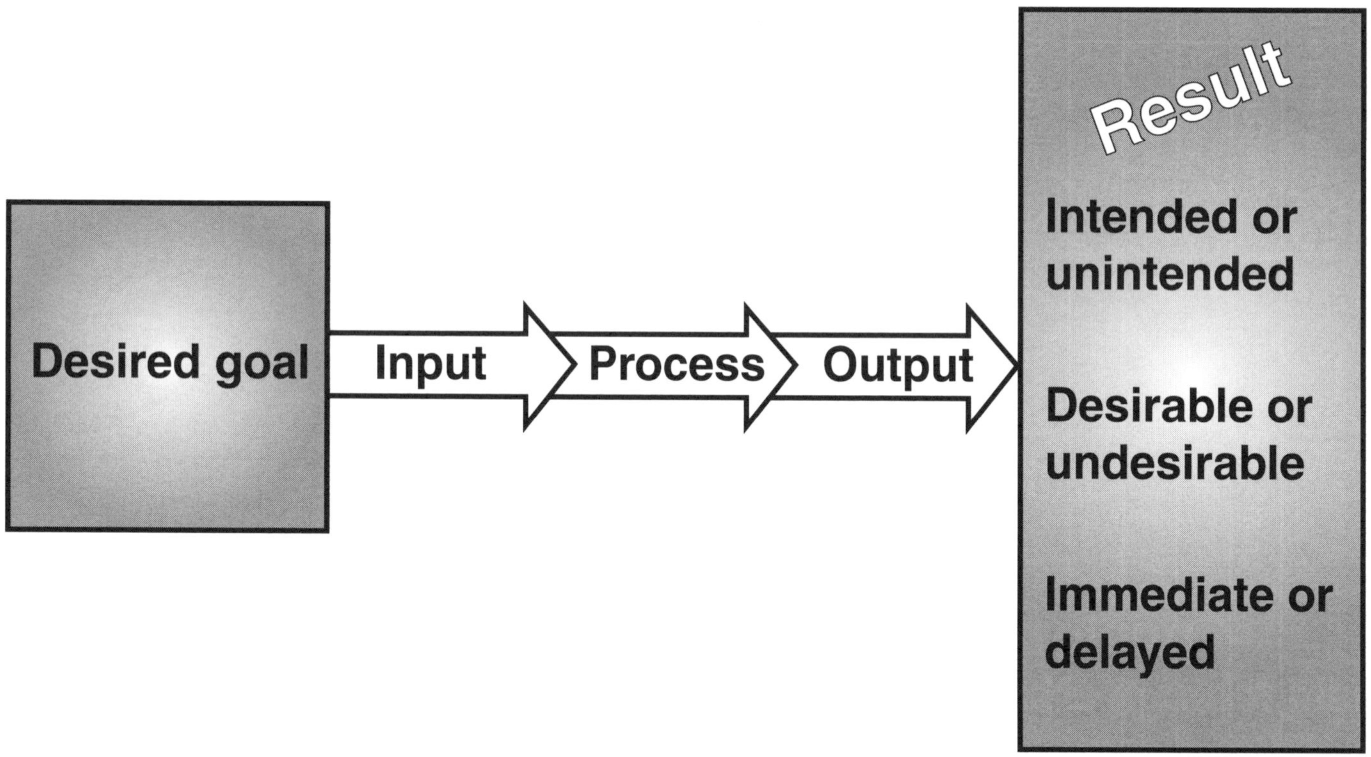

Name ______________________ Date ________

Period ______________________ Score ________

Systems

All systems have the same basic parts, even technology systems such as production, communication, and transportation. Generally, there are five basic elements in a system:

- ❑ **Goals.** The reasons for developing and operating the system.
- ❑ **Inputs.** The material resources or raw materials.
- ❑ **Processes.** The actions taking place.
- ❑ **Outputs.** The results of the inputs and the processes or activities.
- ❑ **Feedback.** Information that modifies or stops the processes.

The drawing below shows a furnace used to heat a building. Look at the drawing. Study it. Write the function of each part of the furnace in the blanks. Label an input, a process, an output, and a feedback.

A. ______________________________________

B. ______________________________________

C. ______________________________________

D. ______________________________________

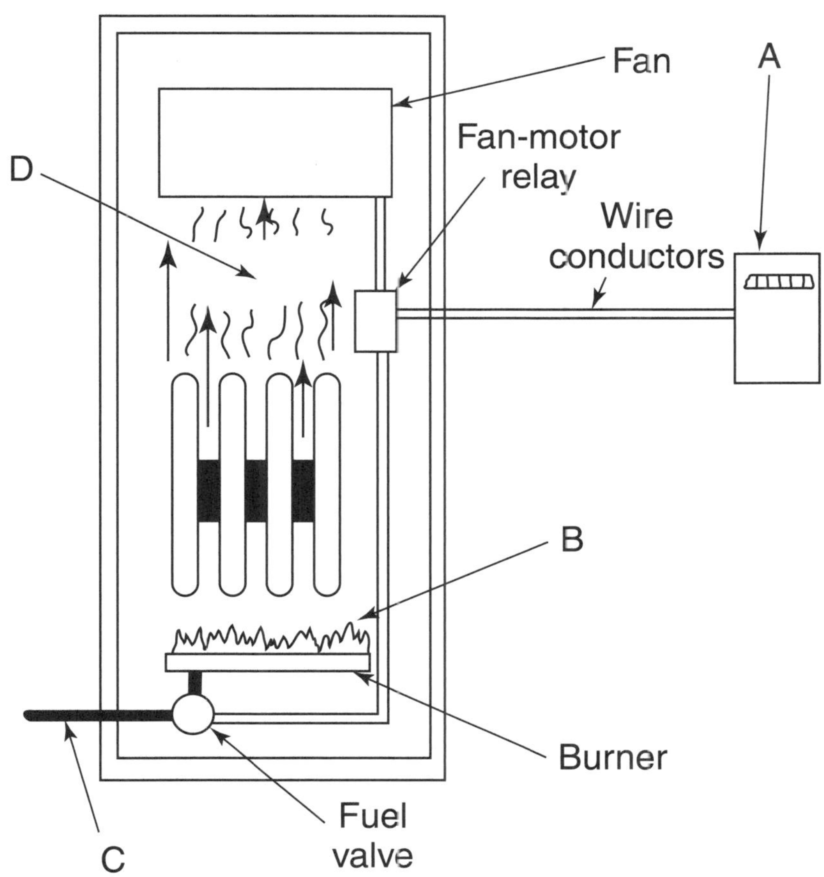

Reproducible Master 2-3

Section 2
Resources and Technology

Focus: Exploration of the inputs to technological systems.

- ❑ Chapter 4: Tools and Technology
- ❑ Chapter 5: Materials and Technology
- ❑ Chapter 6: Energy and Technology
- ❑ Chapter 7: Information and Technology
- ❑ Chapter 8: People, Time, Money, and Technology

Teaching Materials

Text

- Chapter 4, pages 78–107
  - Chapter Activities, pages 99–100
  - Test Your Knowledge, page 100
  - Chapter 4 Activity A—Tools and Materials as Resources, pages 101–106
  - Chapter 4 Activity B—Tools as a Resource, page 107
- Chapter 5, pages 108–135
  - Chapter Activities, page 129
  - Test Your Knowledge, pages 129–130
  - Chapter 5 Activity A—Materials as a Resource, pages 131–135
- Chapter 6, pages 136–163
  - Chapter Activities, page 155
  - Test Your Knowledge, page 155
  - Chapter 6 Activity A—Energy as a Resource, pages 156–162
  - Chapter 6 Activity B—Energy and Machines as Resources, page 163
- Chapter 7, pages 164–181
  - Chapter Activities, page 176
  - Test Your Knowledge, page 176
  - Chapter 7 Activity A—Information as a Resource, pages 177–181
- Chapter 8, pages 182–201
  - Chapter Activities, page 199
  - Test Your Knowledge, page 200

Tech Lab Workbook

- Chapter 4
  - Chapter 4 Activity A Worksheet—Tools and Materials as Resources, pages 21–23
  - Chapter 4 Activity B Worksheet—Tools as a Resource, page 25
  - Chapter 4 Study Questions, pages 27–29
- Chapter 5
  - Chapter 5 Activity A Worksheet—Materials as a Resource, pages 31–32
  - Student Activity 5-1—Product Analysis, pages 33–34
  - Chapter 5 Study Questions, pages 35–36
- Chapter 6
  - Chapter 6 Activity A Worksheet—Energy as a Resource, page 37
  - Student Activity 6-1—Tools and Energy as Resources, page 39
  - Chapter 6 Study Questions, page 41
- Chapter 7
  - Chapter 7 Activity A Worksheet—Information as a Resource, pages 43–44
  - Student Activity 7-1—Developing an Information Chart, page 45
  - Chapter 7 Study Questions, page 47
- Chapter 8
  - Chapter 8 Study Questions, page 49

Teacher's Resources

- Chapter 4
  - Reproducible Master 4-1—Measurement, page 75
  - Reproducible Master 4-2—Basic Mechanisms, page 76
  - Reproducible Master 4-3—From Tool to Mechanism to Machine, page 77
  - Reproducible Master 4-4—Tool Relationships, page 78

- Reproducible Master 4-5—Tool-Usage Chart, page 79
- Reproducible Master 4-6—Tool Safety, page 80

Chapter 5
- Alternative Activity—Testing Material Hardness, pages 71–72
- Reproducible Master 5-1—Properties of Materials, page 81
- Lesson Slide 5-1—Extracting Natural Resources (CD only)
- Lesson Slide 5-2—Producing Industrial Materials (CD only)

Chapter 6
- Reproducible Master 6-1—Nature's Energy Storehouse, page 82
- Reproducible Master 6-2—How a Nuclear Power Station Operates, page 83
- Reproducible Master 6-3—Power Systems in Your Community and Their Energy Sources, page 84
- Lesson Slide 6-1—Energy Use (CD only)
- Lesson Slide 6-2—Types of Energy (CD only)
- Lesson Slide 6-3—Forms of Energy (CD only)

Chapter 7
- Alternative Activity—Evolution of Information Processing, pages 73–74
- Reproducible Master 7-1—Data, Information, and Knowledge, page 85
- Reproducible Master 7-2—Technological Knowledge, page 86

Chapter 8
- Reproducible Master 8-1—Management, page 87
- Reproducible Master 8-2—Functions of Management, page 88
- Reproducible Master 8-3—Levels of Authority and Responsibility, page 89
- Reproducible Master 8-4—Preparing a Company Organizational Chart, page 90
- Reproducible Master 8-5—Preparing an Organizational Chart for Your School, page 91

Introduction

All technological systems use resources to extend human potential. Technological systems use tools, mechanisms, and machines to grow and harvest crops, manufacture products, convert and use energy, construct structures, heal illnesses, communicate information, and transport people and goods. Using human-made artifacts separates technological systems from the natural systems studied in science classes.

Objectives of the Section

At the end of Section 2, each student should be able to do the following:

- ❑ Summarize the roles of humans as toolmakers and users.
- ❑ Compare a tool, mechanism, and machine.
- ❑ Give examples of the major types of tools humans use to extend their potentials.
- ❑ Summarize the characteristics of the major types of machines.
- ❑ Explain natural materials.
- ❑ Explain the major types of engineering materials.
- ❑ Summarize the difference between renewable and exhaustible materials.
- ❑ Summarize several energy sources.
- ❑ Give examples of the major forms of energy.
- ❑ Understand the definition of *energy conversion*.
- ❑ Explain the application and conservation of energy.
- ❑ Apply the use of symbols and signs in communication information.
- ❑ Recall the types of jobs available in technological systems.
- ❑ Summarize the role of money in operating technological systems.

Suggested Schedule and Length of the Section (10–20 Days)

Section 2 introduces students to the resources (inputs) of technological systems. Have them complete the following assignments:

1. Read Chapters 4–8 in their textbooks and answer associated study questions.
2. Participate in several discussions and presentations.
3. Complete two or more laboratory activities and follow-up sheets. These activities will reinforce the study of the following topics:
   A. The use of tools and machines.
   B. The types and properties of materials.
   C. The role of energy in technological systems.
   D. Information in technological systems.
   E. The opportunities for people to work in and use technological systems.

Schedule for Section 2— 18-Week and 36-Week* Classes	
Day 1	Discussion: "Resources and technology." Assign Chapter 4: "Tools and Technology" to be read outside of class. Introduce first activity (tools or materials activity).
Day 2	Discussion: "Tools and technology." Work on activity.
Day 3–4	Work on activity. Assign Chapter 5: "Materials and Technology" to be read outside of class. Discussion: "Materials and technology."
Day 5	Complete and summarize activity. Assign Chapter 6: "Energy and Technology" to be read outside of class.
Day 6	Discussion: "Energy and technology." Introduce second activity (energy activity).
Day 7–9	Work on activity. Assign Chapter 7: "Information and Technology" to be read outside of class.
Day 10	Discussion: "Information and technology." Complete and summarize activity. Assign Chapter 8: "People, Time, Money, and Technology" to be read outside of class.
Day 11	Discussion: "People, time, money, and technology." Introduce third activity (information or people activity).
Day 12–14	Work on activity.
Day 15	Complete and summarize activity. Review and summarize section.

* For a 36-week class, add another activity and spread the discussions over the four weeks scheduled for this section.

Preparing to Teach the Section

Before introducing this section to the students, complete the following tasks:

1. Study this section of the Teacher's Resource.
2. Read Chapters 4–8 in the textbook to review the content.
3. Prepare 10- to 15-minute presentations and discussions on the following topics:
   A. Types of tools. (Textbook, pages 79–89)
   B. Mechanisms and machines. (Textbook, pages 89–97)
   C. Types and properties of materials and obtaining materials. (Textbook, Chapter 5)
   D. Nature, sources, and supplies of energy. (Textbook, Chapter 6)
   E. Information and technology. (Textbook, Chapter 7)
   F. People, time, and money in technology. (Textbook, Chapter 8)
   G. Resources and technology—A summary.
4. Select appropriate laboratory activities from the following options:
   A. Tools and Materials as Resources (Chapter 4), Textbook, pages 101–106.
   B. Tools as a Resource (Chapter 4), Textbook, page 107.
   C. Materials as a Resource (Chapter 5), Textbook, pages 131–135.
   D. Product Analysis (Chapter 5), Tech Lab Workbook, page 33.
   E. Testing Material Hardness (Chapter 5), Teacher's Resource, page 71.
   F. Energy as a Resource (Chapter 6), Textbook, pages 156–162.

G. Energy and Machines as Resources (Chapter 6), Textbook, page 163.
H. Tools and Energy as Resources (Chapter 6), Tech Lab Workbook, page 39.
I. Information as a Resource (Chapter 7), Textbook, pages 177–181.
J. Developing an Information Chart (Chapter 7), Tech Lab Workbook, page 45.
K. Evolution of Information Processing (Chapter 7), Teacher's Resource, page 73.

5. Make necessary material and supply preparations for presenting the laboratory activities.
6. Work through the laboratory activities and student study questions presented in the textbook, Tech Lab Workbook, and Teacher's Resource.

Presenting the Section

This section includes time for short, informative discussions and presentations supported by meaningful laboratory activities. The discussions should be limited to 15 minutes and reinforce the information in the textbook. Summarize all discussion and laboratory work through the use of the Test Your Knowledge questions accompanying the textbook chapters and study questions in the Tech Lab Workbook. On the last day of the section, summarize the concepts that have been presented. Direct the students' attention to what they have learned over the past two to four weeks.

Adjustments for Students with Special Needs and Gifted Students

The 13 activities suggested for this section are designed to introduce the primary resource inputs to technological systems. They lend themselves to either individual or group work. One way to adjust instruction for differing abilities and varying interests is to assign the elected activities to groups with a mix of students. This can group gifted students, average students, and students with special needs, as well as those who think logically and sequentially with those who are random thinkers. Special attention and individual help should be given to those students who require more structure. If the activity is highly structured for all students, however, the gifted students might lack the level of challenge needed for them to do their best work.

Evaluation Procedures

The emphasis of this section is on understanding how resources are used within a technological system. The focus should be on the importance of tools, materials, energy, information, people, time, and money in implementing technology. Specific tool nomenclature, craftsmanship of the final products, memorization of material properties, and other similar criteria are of secondary importance. The goal is to understand that materials have a range of properties that must be evaluated when developing products. Therefore, the students should be evaluated on the following criteria:

1. Contributions to group activity. This evaluation is possible with daily logs (diaries) the students maintain, teacher observations, and student feedback.
2. Laboratory sheets, sketches, and designs completed during the activities.
3. Participation in discussions conducted during the laboratory activities.
4. Scores on quizzes and chapter questions.

Implementing the Laboratory Activities

A number of different activities have been developed for this section. These can be found at the end of Section 2 (Chapters 4–8) of the textbook, in Section 2 of the Tech Lab Workbook, and in this Teacher's Resource. The activities are suggested activities. Any activity introducing the concept that tools, materials, information, energy, people, time, and money are essential for technology can be substituted.

The time constraints for this section restrict the program to a few relatively short activities.

Procedures for Directing Each Suggested Activity

Chapter 4: Tools and Materials as Resources, Textbook, pages 101–106, and Tech Lab Workbook, pages 21–23

This is a prescriptive activity introducing tools as an essential input for technology. The activity presents a specific set of procedures to the students. The system used to make the product should be demonstrated. Divide the class into three groups:

❑ Group 1 makes the game board.
❑ Group 2 makes and stains the pegs.
❑ Group 3 makes the storage-hole cover.

Have each group complete the following steps:

1. Carefully read the directions for making the product.
2. Assign a step or several steps to each member of the group.
3. Obtain the supplies needed to make the product (listed in the textbook).
4. Produce 24 parts, using the procedures listed in the textbook.
5. Analyze the output (quality control).

When all parts are produced, give each group eight sets of parts. Have each group assemble the products by completing the following steps:

1. Carefully read the directions for the assembly procedure.
2. Assign a step or several steps to each member of the group.
3. Assemble eight products, using the assembly procedure.

Chapter 4: Tools as a Resource, Textbook, page 107, and Tech Lab Workbook, page 25

Chapter 4 Activity B is designed to be an open-ended, problem-solving activity. The students should be provided the design brief and encouraged to develop a unique solution to the problem. The activity can be done individually. "Tools as a Resource" will work better, however, with groups of two to four students. Have the students complete the following steps:

1. Define the problem they are attempting to solve.
2. List any restrictions they face, such as marble size and cost of the solution.
3. Develop sketches of two to five ways to solve the problem.
4. Select and refine the best solution.
5. If time permits, build a prototype of the solution.

Chapter 5: Materials as a Resource, Textbook, pages 131–135, and Tech Lab Workbook, pages 31–32

Chapter 5 Activity A is a prescriptive activity to introduce students to material properties. A set of material samples should be provided, containing a variety of materials, such as sheet aluminum, copper, steel, acrylic plastic, plywood, and lumber. Divide the class into groups of two to four students. Have each group complete the following steps:

1. Perform all the appropriate tests the procedure in the textbook outlines.
2. Record test results on the form supplied on page 31 of the Tech Lab Workbook.
3. Present the group's test findings to the class.
4. Discuss the implications of the data for material selection and application.

Chapter 5: Product Analysis, Tech Lab Workbook, page 33–34

This activity is designed to allow the students to analyze products in terms of the materials they contain. To use this activity, divide the class into groups of two to four students. Have each group follow the directions at the top of the activity sheet for Activity 5-1. The groups should complete the following steps:

1. Select a product in the technology-education laboratory.
2. Identify four materials the product contains.
3. List the materials and their uses on page 33 of the Tech Lab Workbook.
4. Indicate possible reasons for selecting each material. These can include reasons such as appearance, cost, ease of manufacture, and strength.
5. Present their findings and encourage the class to react to their reports.

Chapter 5: Testing Material Hardness, Teacher's Resource, pages 71–72

"Testing Material Hardness" is a prescriptive activity in which the students determine the relative hardness of three selected woods. They will follow a set procedure. Divide the class into groups of two to three students. Provide each group with a hardness tester you have built. (The actual construction of the tester is open to modification.) The key parts are the following:

- ❑ A 3/8″-diameter × 36″ steel rod that has one end rounded.
- ❑ A polyvinyl chloride (PVC) pipe with a 5/32″-diameter hole drilled 2″ from one end. The pipe should have a diameter allowing the steel rod to slip through it easily.
- ❑ A 1/8″ retainer pin that keeps the rod from dropping down the PVC pipe before the test is conducted.
- ❑ A 24″ space between the retainer pin and specimen.
- ❑ Parallel strips on the tester base to ensure the specimen is located directly below the steel rod.

Have each group conduct the tests by completing the following steps:

1. Carefully read the directions for the procedure.
2. Complete a face-grain and end-grain test for each of the three specimens.
3. Record the data.
4. Draw conclusions about the following:
   - ❑ Relative hardness of end grain and face grain of the same species of wood.
   - ❑ Relative hardness of the three species of wood.

Chapter 6: Energy as a Resource, Textbook, pages 156–162, and Tech Lab Workbook, page 37

This is a prescriptive activity allowing the students to make a common energy converter—an electric motor. The activity presents a specific set of procedures to the students. "Energy as a Resource" can be done individually or by groups of students. If using groups, assign three students to each group. Have the students complete the following activities:

- ❑ Student 1—Lay out and produce the base.
- ❑ Student 2—Make the armature.
- ❑ Student 3—Make two motor bearings.
- ❑ All—Assemble and test the motor.

Chapter 6: Energy and Machines as Resources, Textbook, page 163

Chapter 6 Activity B is designed to allow the students to use their creativity to design a system using a simple machine and human strength. There is no correct answer. To use this activity, divide the class into groups of two to four students. Have each group complete the following steps:

1. Define the problem the group is attempting to solve.
2. List any restrictions the group faces, such as materials that can be used and the strength of the person using the machine.
3. Develop a sketch of two to five ways to solve the problem.
4. Select and refine the best solution.
5. Build a prototype of the solution, using only the materials listed in the textbook. (Limiting the materials introduces a common restriction all product designers face.)
6. Demonstrate the solution.
7. List the ways the group can improve its designs.

Chapter 6: Tools and Energy as Resources, Tech Lab Workbook, page 39

This activity is designed to be less structured than the game activity provided in the textbook. The students should be provided the drawings for the rubber band car and challenged to complete the following steps:

1. Develop a bill of materials for the product.
2. Prepare a procedure sheet for building the product, using the form on page 25 of the Tech Lab Workbook.
3. List any restrictions they face.
4. Build a prototype of the racer.
5. Discuss the procedures they used, ways to build the product more efficiently, and other similar topics.

Chapter 7: Information as a Resource, Textbook, pages 177–181, and Tech Lab Workbook, pages 43–44

Chapter 7 Activity A is designed to allow students to read, review, and synthesize information. Generally, this activity is done individually. The students can, however, work in groups of two. To use the activity, have the students complete the four steps outlined on page 177 of the textbook. As an alternative, this activity can be assigned as homework.

Chapter 7: Developing an Information Chart, Tech Lab Workbook, page 45

This is a prescriptive activity, designed to allow students to analyze sources and types of information. They will locate written information (procedures), graphic information (such as sketches and drawings), and charts and tables (bill of materials). The students should follow the directions contained at the top of page 45 in the Tech Lab Workbook.

An alternate activity is to give the students a simple product and have them develop the three types of information. The product can be a package; a baked good, such as a cookie; or some other simple item. Another alternate activity is to have the students develop pictorial directions for baking a cake. Each step should be shown using only pictures. Quantities can be shown on a measuring cup or spoon. All other information, however, must be shown in drawings. The students should use the form provided on page 45 of their Tech Lab Workbooks.

Chapter 7: Evolution of Information Processing, Teacher's Resource, pages 73–74

This is an open-ended activity in which students practice and compare five methods of recording information that have been developed over time:

- ❑ Clay tablets.
- ❑ Pen, ink, and paper.
- ❑ Movable type.
- ❑ Typewriting.
- ❑ Information processing.

The students will first record a message and then be asked to change it slightly. They will compare and contrast each method in terms of the following criteria:

- ❑ Ease of producing the message.
- ❑ Ease of revising the message.

Procedure

1. Divide the class into groups of five students.
2. Have each member of the group draw at random a method for producing a message:
   - ❑ Clay tablet.
   - ❑ Pen, ink, and paper.
   - ❑ Movable type.
   - ❑ Typewriter.
   - ❑ Word processor.
3. Have each member of the group gather the supplies and equipment needed to produce the message.
4. Have the students produce the following message using their technology: **Technology education is a new basic.**
5. Have the students change their messages to read as follows: **Technology education is THE new basic.**

6. Have everyone discuss with their group members the following topics:
    - ❑ Ages of the technologies they each used.
    - ❑ Methods of producing the messages.
    - ❑ Ease of producing the original messages.
    - ❑ Methods of changing the messages.
    - ❑ Ease of changing the messages.

Answer Key for Section 2

Chapter 4—Tools and Technology

Test Your Knowledge, Textbook, page 100

1. D. All of the above.
2. tools
3. mechanism
4. Answers will vary. Some acceptable answers are a ruler (measuring), handsaw (cutting), spade bit (drilling), vise (gripping), claw hammer (pounding), and sharpening stone (polishing).
5. distance multiplier
6. True
7. A lever is a device consisting of a lever arm and a fulcrum. This device can increase the force applied to the work or the amount of movement created. A wheel and axle consists of a shaft attached to the center of a disc. This mechanism operates as a second-class lever. A pulley is a grooved rim attached to a loose axle. This mechanism can act to change the direction of force, multiply force, or multiply distance. An inclined plane is a sloped surface. This mechanism operates on the principle that moving up a slope is easier than lifting straight up. A wedge is a set of two inclined planes. A screw is an inclined plane wrapped on a round shaft.
8. A. Frame., B. Tool., C. Electric motor., and D. Feed control.
9. Turning machines use a stationary tool to cut a material rotating around an axis. Drilling machines rotate and push downward to cut a hole into a work. Milling and sawing machines have motors that cause blades to feed into a work in order to create a cut. Planing and shaping machines use single-point tools to produce flat cuts on the surface of a work. Grinding and sanding machines have moving abrasives. These abrasives cut materials from the workpiece. Shearing machines use opposed edges to slice materials into parts.
10. Proper tool storage is important to keep people from being injured from sharp points and edges and to protect from the toll from the environment (such as rusting and corrosion).

Study Questions, Tech Lab Workbook, page 27–29

1. D. All of the above.
2. task
3. B. Tool., C. Mechanism., A. Machine.
4. Primary tools are the older tools people use to make new tools. They are necessary because, without them, we could not develop new tools and machines to make lives easier.
5. A. Load.
   B. Fulcrum.
   C. Force.
6. first-class
7. B. distance
8. A. Lever.
   B. Wheel and axle.
   C. Pulley.
   D. Inclined plane.
   E. Wedge.
   F. Screw.
9. Basic structure, power unit, control unit, tool.
10. Turning machines, drilling machines, milling and sawing machines, shaping and planing machines, grinding and sanding machines, and shearing machines.
11. tool
12. owner's manual

Chapter 5—Materials and Technology

Test Your Knowledge, Textbook, page 129–130

1. A substance appearing in nature out of which a useful item can be made.
2. B. Trees., C. Rocks., and D. Petroleum.
3. Answers will vary. Some acceptable answers are coal, petroleum, and natural gas.
4. engineering materials
5. B. Metal., C. Glass., D. Plastic., and E. Concrete.
6. renewable
7. exhaustible resources
8. Producing plastic.—B. Chemical processing. Making lumber.—A. Mechanical processing. Producing steel.—C. Thermal processing. Examples will vary. Some acceptable answers are removing gold from its ore (chemical), grinding grain into flour (mechanical), and making glass (thermal).
9. A material that results from primary processing, but needs further processing to become part of consumer products.
10. Examples include flour, gravel, lumber, steel, glass, and aluminum.
11. A characteristic of a material.
12. Physical properties describe the basic features of a material. Mechanical properties affect how a material reacts to mechanical force and loads. Chemical properties describe a material's reaction to chemicals. Thermal properties are a material's reaction to heat. Electrical and magnetic properties describe a material's reaction to electrical current and magnetic forces. Optical properties govern a material's reaction to light. Acoustical properties determine a material's reaction to sound waves.
13. The materials can become contaminated, rusted, corroded, rotten, or deformed.

Study Questions, Tech Lab Workbook, page 35–36

1. natural resources
2. engineering materials
3. They are all categories of engineering materials.
4. C. inorganic materials
5. C. ceramics
6. True
7. A. renewable
   B. exhaustible
   C. renewable
   D. renewable
   E. exhaustible
   F. exhaustible
   G. exhaustible
8. drilling wells and digging mines
9. mechanical
10. thermal processing
11. industrial material
12. Physical, mechanical, chemical, thermal, electrical and magnetic, optical, and acoustical.
13. mechanical properties

Chapter 6—Energy and Technology

Test Your Knowledge, Textbook, page 155

1. Answers will vary. Some acceptable answers include the following responses: In natural systems, energy is used to do work and accomplish daily activities. In technological systems, energy is used to power cars and machines, light homes, treat illnesses, plant crops, heat buildings, cook food, and provide entertainment.
2. energy
3. work
4. power
5. Examples include heat in a lightbulb and friction in a motor.
6. A. Radiant energy.
   B. Chemical energy.
   C. Electrical energy.
   D. Mechanical energy.
   E. Heat energy.
   F. Nuclear energy.
7. Primary energy resources are in the form of natural resources. Secondary energy resources are in the form of altered, more usable resources.
8. Peat, coal, petroleum, and natural gas.
9. hydropower
10. geothermal energy
11. bioenergy

Study Questions, Tech Lab Workbook, page 41

1. True
2. work
3. potential energy
4. kinetic energy
5. B. different things. Energy is the ability to do work, while power is the speed at which work is done.
6. British thermal unit (Btu)
7. False
8. Radiant, heat, mechanical, chemical, electrical, and nuclear energy.
9. the Sun
10. fossil
11. fission
12. geothermal energy
13. bioenergy

Chapter 7—Information and Technology

Test Your Knowledge, Textbook, page 176

1. scientific
2. descriptive
3. technological
4. information
5. knowledge
6. Everything happening in nature can be understood. Nature is always the same. There is a relationship between a cause and its effect.
7. Technological knowledge is essential because it helps us do things, make products and systems, and control and modify the natural and human-made environments. Without this knowledge, we could not feed everyone in the world, process materials into usable products, or harness energy to create power.
8. Answers will vary. Some acceptable answers are radio, television, libraries, books, magazines, the Internet, and other people.
9. descriptive research
10. experimental research

Study Questions, Tech Lab Workbook, page 47

1. A. Humanities knowledge.
   B. Technological knowledge.
   C. Scientific knowledge.
   D. Descriptive knowledge.
   E. Humanities knowledge.
   F. Technological knowledge.
2. data
3. information
4. knowledge
5. False
6. Information
7. Historical, descriptive, and experimental research.
8. historical research

Chapter 8—People, Time, Money, and Technology

Test Your Knowledge, Textbook, page 200

1. The kinds of work change because advancements and discoveries are made. With these changes come new demands.
2. Has a high degree of skill in a narrow range.—B. Semiskilled
   Requires little or no training.—C. Unskilled
   Does routine work.—C. Unskilled
   Might serve an apprenticeship.—A. Skilled
3. Ideas and knowledge.
4. economic; resources; product, service, or system; profit
5. Serves as a major link between engineers and the factory floor.—C. Technologist
   Organizes resources to produce products.—A. Manager
   Designs products, systems, and structures.—B. Engineer
   Solves operational problems.—D. Technician
   Often leads a design team.—B. Engineer
   Constructs products or structures.—E. Worker
6. The factors are personal goals, interests, abilities, and physical traits.
7. career ladder

8. True
9. False
10. True

Study Questions, Tech Lab Workbook, page 49

1. white-collar workers
2. Skilled, semiskilled, and unskilled workers.
3. ideas, knowledge
4. profit
5. managers
6. Engineers
7. technologists
8. talents
9. career ladder
10. Time
11. Money

Captions for Reproducible Masters

- ❑ **Reproducible Master 4-1.** Measurement—Measuring tools are important technological resources.
- ❑ **Reproducible Master 4-2.** Basic Mechanisms—The six basic mechanisms can be explained by two main principles.
- ❑ **Reproducible Master 4-3.** From Tool to Mechanism to Machine—Machines are combinations of tools and mechanisms.
- ❑ **Reproducible Master 4-4.** Tool Relationships—A chart for students to fill out while learning that machines are made up of tools and mechanisms.
- ❑ **Reproducible Master 4-5.** Tool-Usage Chart—A chart for students to fill out while studying the procedures for using certain tools.
- ❑ **Reproducible Master 4-6.** Tool Safety—A chart for students to fill out while learning safety rules for using certain tools.
- ❑ **Reproducible Master 5-1.** Properties of Materials—There are seven major categories of material properties.
- ❑ **Reproducible Master 6-1.** Nature's Energy Storehouse—Plants store energy from sunlight so animals can use it.
- ❑ **Reproducible Master 6-2.** How a Nuclear Power Station Operates—A diagram showing how fuel is turned into electricity at a nuclear power station.
- ❑ **Reproducible Master 6-3.** Power Systems in Your Community and Their Energy Sources—A chart for students to fill out when discussing power systems and sources of energy.
- ❑ **Reproducible Master 7-1.** Data, Information, and Knowledge—The differences among data, information, and knowledge are stated.
- ❑ **Reproducible Master 7-2.** Technological Knowledge—Technological knowledge is the knowledge of action.
- ❑ **Reproducible Master 8-1.** Management—A company's management guides and directs the company's activities.
- ❑ **Reproducible Master 8-2.** Functions of Management—The main functions of management are to plan, organize, actuate, and control.
- ❑ **Reproducible Master 8-3.** Levels of Authority and Responsibility—There are typically three levels of management.
- ❑ **Reproducible Master 8-4.** Preparing a Company Organizational Chart—A worksheet for students to complete to practice producing organizational charts.
- ❑ **Reproducible Master 8-5.** Preparing an Organizational Chart for Your School—A worksheet for students to complete for additional practice in the production of organizational charts.

Captions for Lesson Slides (CD Only)

- ❑ **Lesson Slide 5-1.** Extracting Natural Resources—Humans cannot create exhaustible resources. New sources must be found after these resources are extracted.
- ❑ **Lesson Slide 5-2.** Producing Industrial Materials—Most raw materials must be processed into something more

useful. The main categories of primary processing are listed.

- ❑ **Lesson Slide 6-1.** Energy Use—Energy use has changed over time. Today, each person uses more than 100 times the energy early humans used.
- ❑ **Lesson Slide 6-2.** Types of Energy—Energy is either moving or stored. Potential energy is stored energy. Kinetic energy is moving energy.
- ❑ **Lesson Slide 6-3.** Forms of Energy—Energy appears in several different forms.

Name ______________________ Date __________

Period ______________________ Score __________

Chapter 5 Alternative Activity

Testing Material Hardness

The Challenge

Determine the relative hardness of the face and end grain of three selected species of common wood.

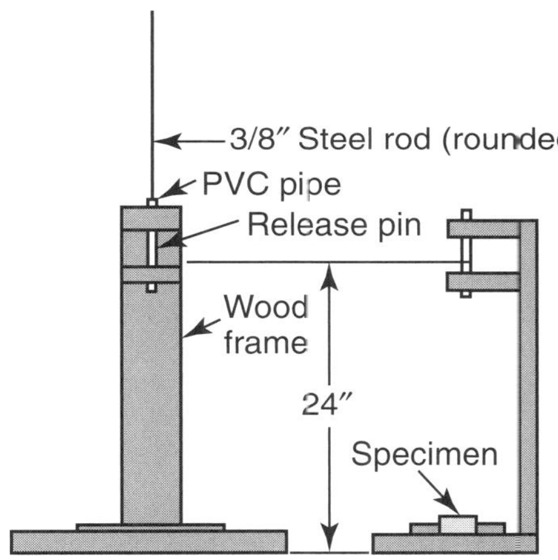

Background

Material properties are important in developing technological devices. One key property is hardness. This is a material's ability to resist scratching and denting. Technologists have a series of tests they can use to determine a material's hardness. Often, these tests measure the size of a dent a test-instrument probe that has a load applied to it produces in a specimen.

Materials

To complete your hardness test, you need the following materials and equipment:

- ❑ One hardness tester.
- ❑ One sample each of three woods (each sample should measure 1 1/2″ × 1 1/2″ × 1 1/2″).
- ❑ One piece of carbon paper.
- ❑ Paper to record results.

Chapter 5 Alternate Activity *(continued)* Name ____________________

Procedure

1. Obtain 1 1/2″ × 1 1/2″ × 1 1/2″ samples of three types of wood.
2. Number the samples.
3. Prepare a recording sheet similar to the one below.

Sample #	Species	End Grain	Face Grain
1			
2			
3			

4. Record the name of the species for each sample.
5. Obtain a hardness tester from your teacher.
6. Raise the steel rod. Place the release pin beneath it.
7. Place specimen #1 face grain up, directly beneath the steel rod.
8. Place a piece of carbon paper over the specimen.
9. Pull the release pin out, allowing the steel rod to drop.
10. Remove the specimen.
11. Measure the diameter of the dent the rod produced.
12. Record the measurement on your data chart.
13. Repeat steps 6–12 for the end grain of specimen #1.
14. Repeat steps 6–13 for specimens #2 and #3.

Additional Challenges

1. Identify another material property you can test, such as the following:
   - ❑ Thermal conductivity.
   - ❑ Tensile strength.
   - ❑ Electrical conductivity.
2. Determine how you might test this property.
3. Develop a test apparatus to test this property.
4. Develop a data-recording sheet for the test.
5. Conduct the test.
6. Report your findings.

Name ______________________ Date __________

Period ______________________ Score __________

Chapter 7 Alternative Activity

Evolution of Information Processing

The Challenge

Compare and contrast ways people have processed and recorded information over time.

Background

Information is a basic input to all technological systems. This input is often processed and recorded for future reference. Historically, the techniques used have changed greatly. Early people made drawings on cave walls and carved messages in stone and clay tablets. Later, handwriting evolved. Scribes used pens and ink to save messages. This technique was changed with the advent of movable type and printing. Even more recently, the typewriter allowed individuals to produce messages with speed and ease. Today, the personal computer (PC) is commonplace in almost all areas of business and personal life.

Materials

For this activity, you need the following supplies.

Clay Tablets

- ❑ Potter's clay or modeling clay.
- ❑ A pointed, 1/4″-diameter dowel, about 6″ long.

Chapter 7 Alternate Activity *(continued)* Name ___________________________

Pen, Ink, and Paper

- ❑ A turkey or chicken feather (quill pen).
- ❑ India ink.
- ❑ Parchment paper.

Movable Type

- ❑ A flexible rubber stamp set (available at office-supply, discount, and toy stores).
- ❑ An ink pad.
- ❑ Bond paper.

Typewriting

- ❑ An electric typewriter or a manual typewriter without an erasing feature.
- ❑ Typing paper.

Information Processing

- ❑ A computer.
- ❑ A printer.
- ❑ Word processing software.
- ❑ Printer paper.

Procedure

1. Draw at random a method for producing a message:
    - ❑ Clay tablet.
    - ❑ Pen, ink, and paper.
    - ❑ Movable type.
    - ❑ Typewriter.
    - ❑ Word processor.
2. Gather the supplies and equipment needed to produce the message.
3. Produce the following message using your technology: **Technology education is a new basic.**
4. Change your message to read as follows: **Technology education is THE new basic.**
5. Discuss with your group members the following topics:
    - ❑ Ages of the technologies they each used.
    - ❑ Methods of producing the messages.
    - ❑ Ease of producing the original messages.
    - ❑ Methods of changing the messages.
    - ❑ Ease of changing the messages.

Additional Challenges

1. Prepare a bulletin board or showcase display showing the evolution of information processing and recording technology.
2. Visit a modern printing or publishing plant (such as a newspaper or commercial printer) to observe information processing and storage technologies.

Measurement

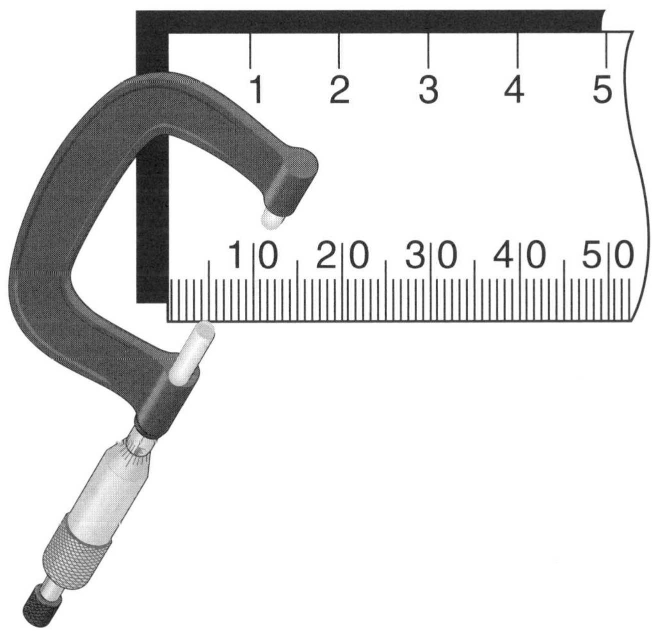

Determining Linear Sizes

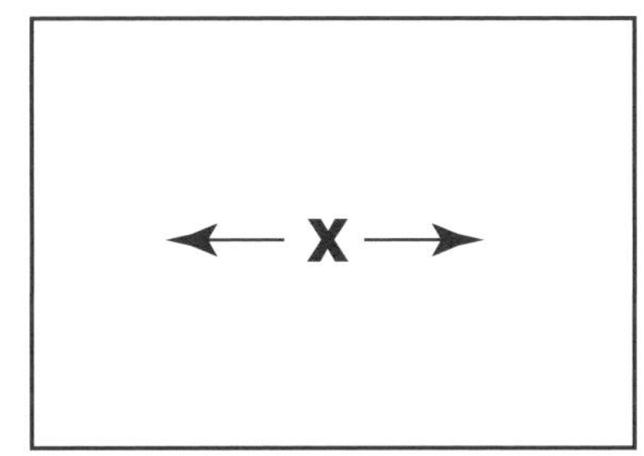

Distances

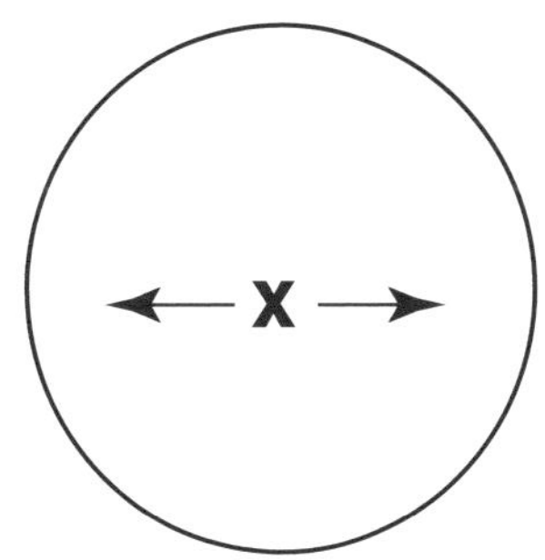

Diameter

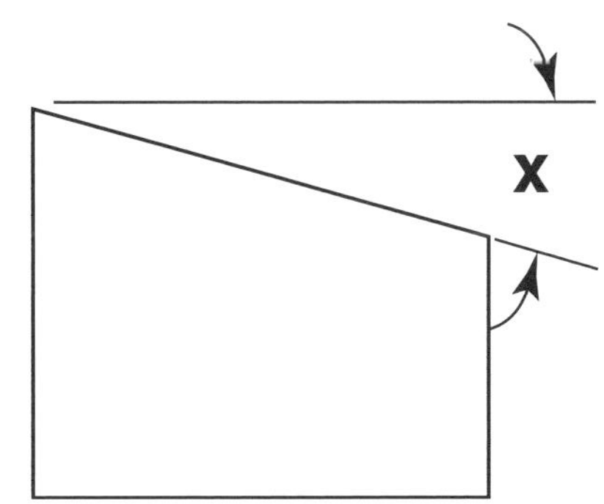

Angles

Reproducible Master 4-1

Basic Mechanisms

Lever Principle
- Lever
- Wheel and axle
- Pulley

Inclined Plane Principle
- Wedge
- Inclined plane
- Screw

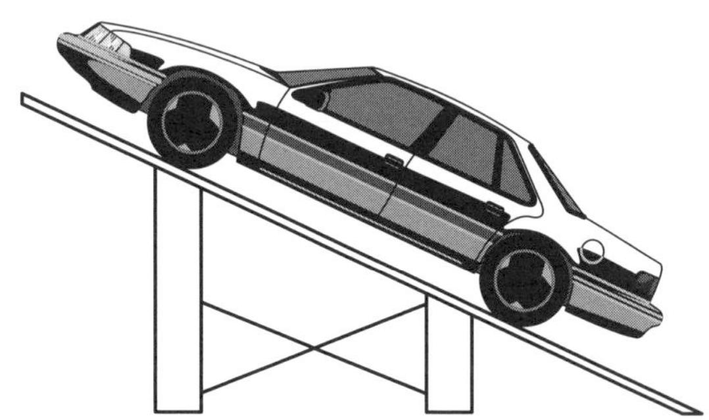

Reproducible Master 4-2

From Tool to Mechanism to Machine

Tool Extension of human hand	**Mechanism** Controls or powers a tool	**Machine** Combination of a tool and a mechanism
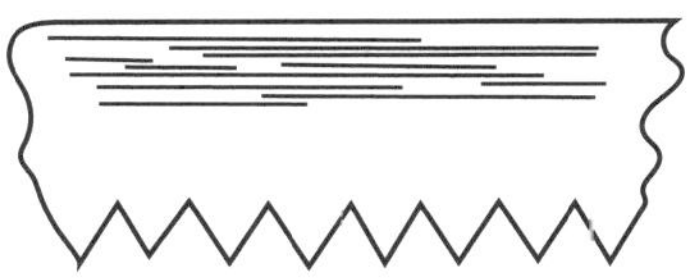 **Saw**	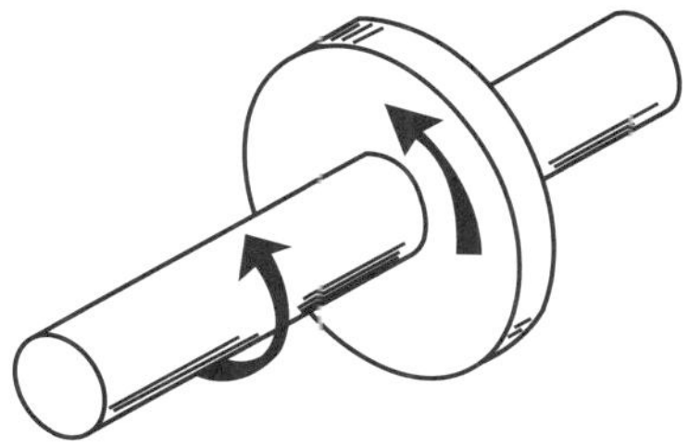 **Wheel and axle**	 **Circular saw**
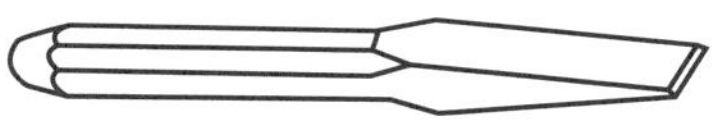 **Chisel**	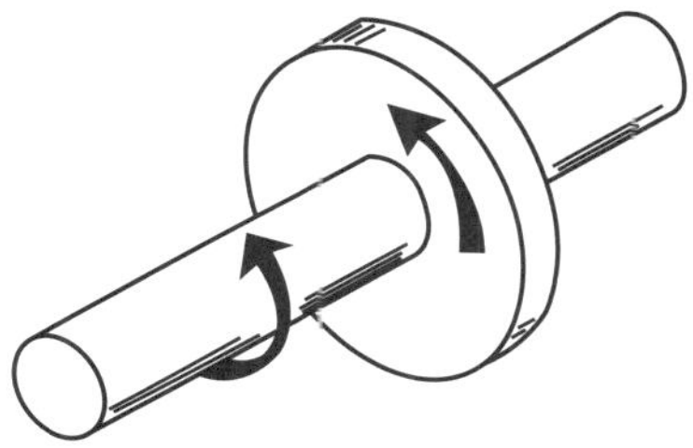 **Wheel and axle**	 **Lathe**
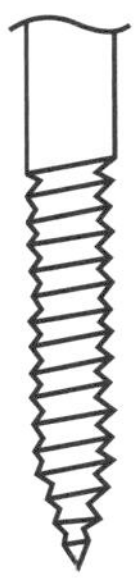 **Screw**	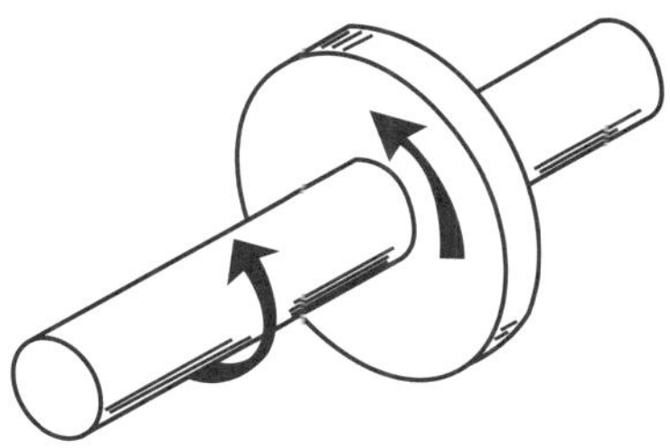 **Wheel and axle**	 **Drill**
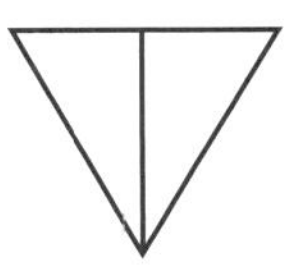 **Wedge**	 **Lever and wedge**	 **Shear**

Reproducible Master 4-3

Name ______________________ Date __________

Period ______________________ Score __________

Tool Relationships

Study the machines in your laboratory. Complete the chart below.

Name of Machine	Tool	Mechanism
Sample: Table saw	Saw	Wheel and axle

Reproducible Master 4-4

Name ______________________ Date__________

Period ______________________ Score __________

Tool-Usage Chart

In the chart below, write the procedure for using each tool you used in developing your product. If other tools were used, list them in the left column. List the procedures for using them in the right column.

Type of Tool	Procedural Steps
Measuring tool	
Cutting tool	
Drilling tool	
Gripping tool	
Pounding tool	
Polishing tool	

Reproducible Master 4-5

Name____________________________ Date__________

Period ___________________________ Score __________

Tool Safety

In the chart below, list the tools that will be used in producing your product in the left column. In the right column, write the safety rules for the use of each tool. (If you do not know a safety rule for a particular tool, check with your teacher. Ask your teacher to demonstrate safe usage of the tool.)

Tools to Be Used	Safety Rules (List for Each Tool)

Reproducible Master 4-6

Properties of Materials

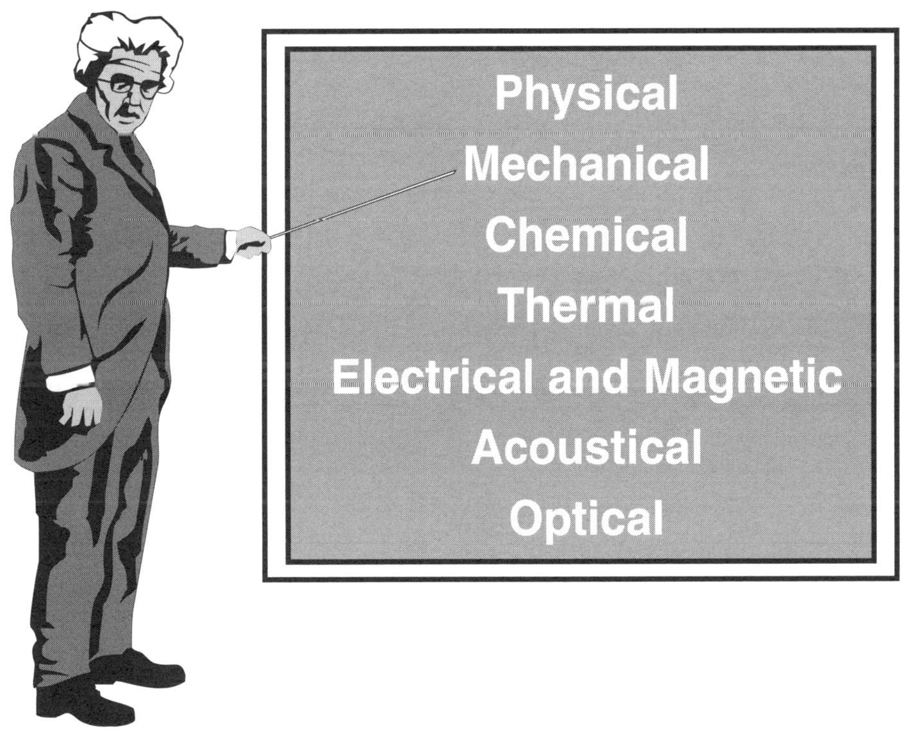

Nature's Energy Storehouse

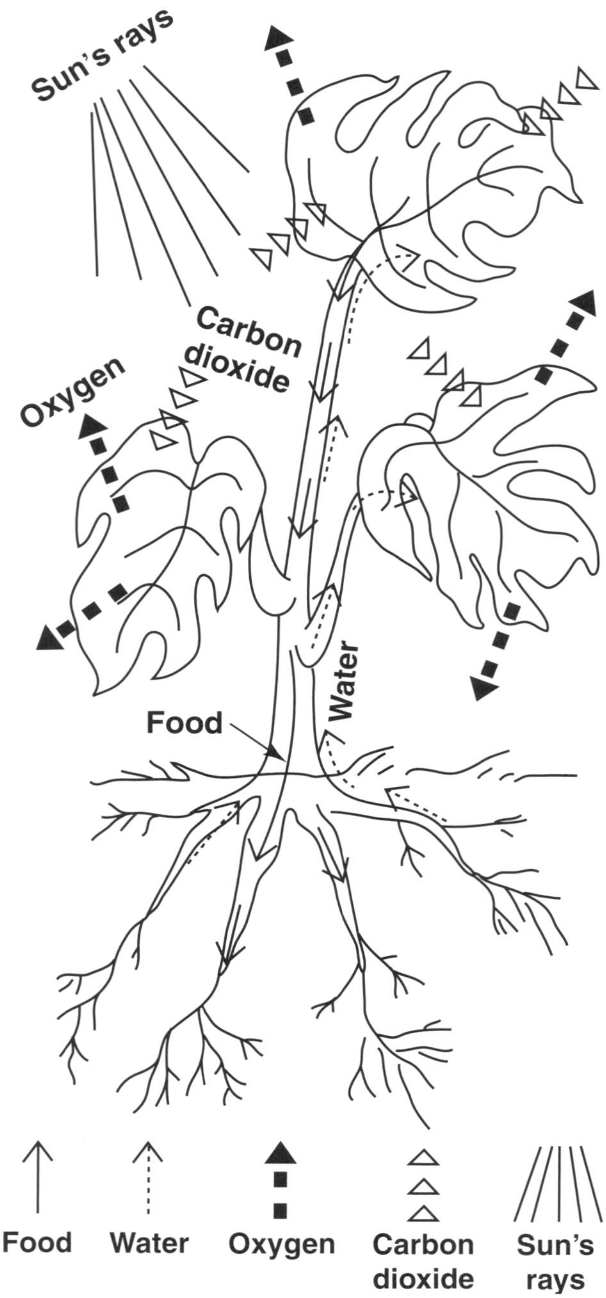

Trees and plant life of all kinds are natural storehouses for energy. They take in sunlight, water, carbon dioxide, and nutrients. Animals derive energy from eating and digesting plant life. Usable energy is the result of plant decay or burning.

Reproducible Master 6-1

How a Nuclear Power Station Operates

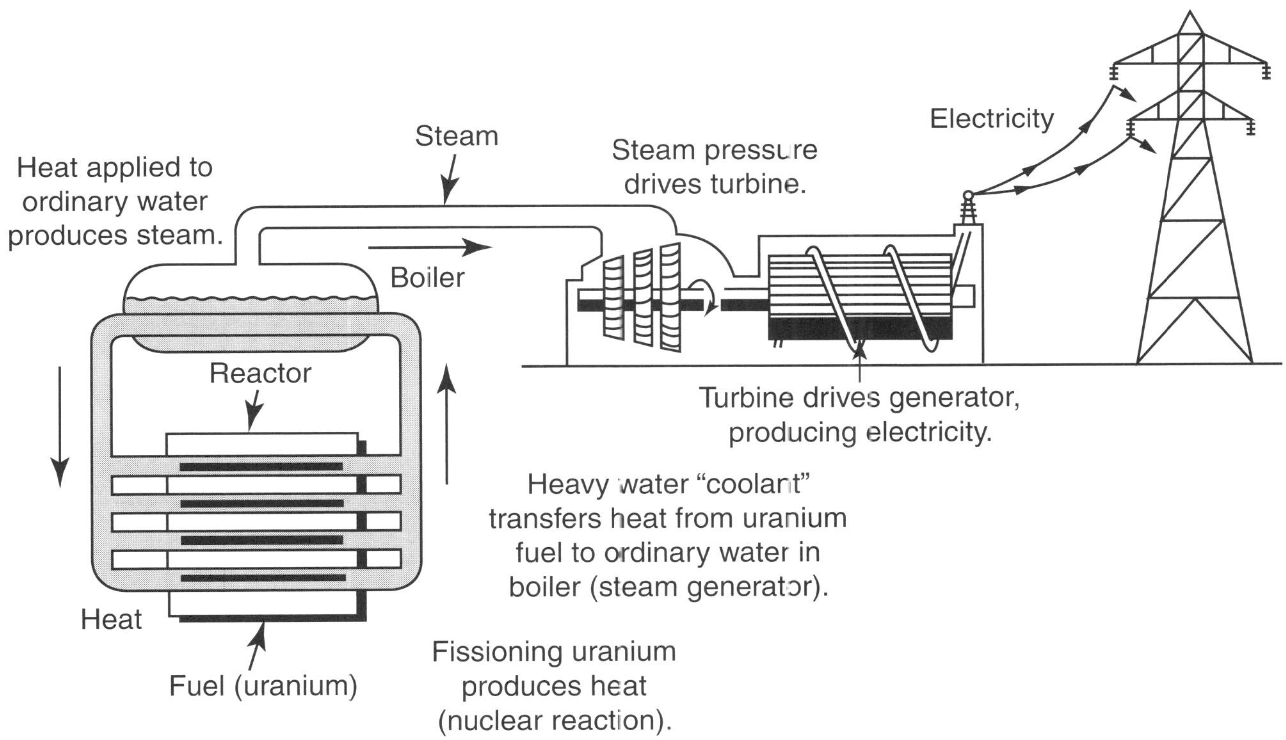

Reproducible Master 6-2

Name ______________________ Date __________

Period _____________________ Score _________

Power Systems in Your Community and Their Energy Sources

Every power system must have a source of power. On the chart below, list all the power systems you can think of in your community. Place the names of the power systems in the left-hand column. List the source of energy each uses in the right-hand column.

Name of Power System	Source of Energy

Reproducible Master 6-3

Data — Information — Knowledge

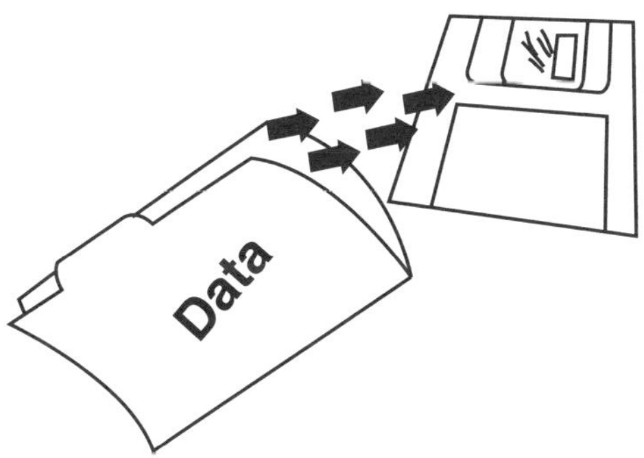

Data — individual facts, statistics (numerical data), and ideas.

Information — sorted and arranged data.

Knowledge — information that humans can apply to situations.

Technological Knowledge

Technological knowledge is the knowledge of action. This knowledge is the knowledge of how to do or how to make something.

Reproducible Master 7-2

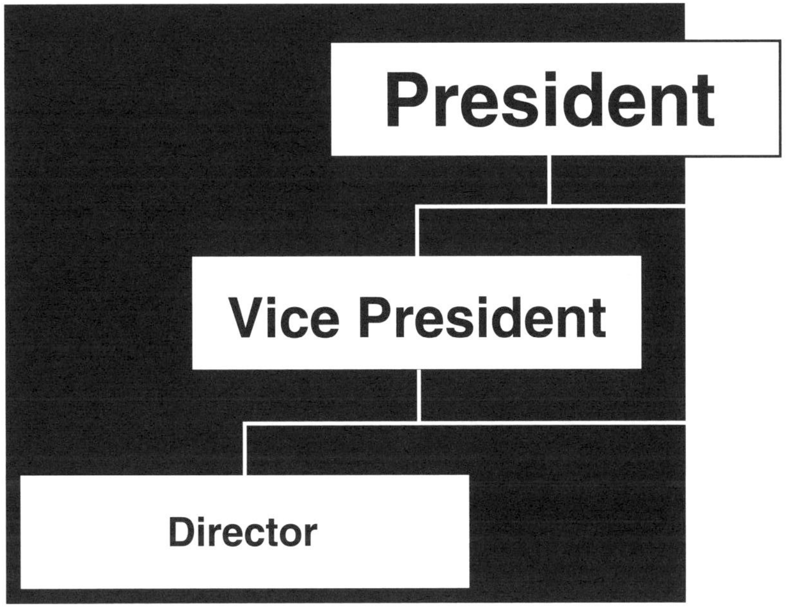

Management

Management guides and directs company activities so they operate in an efficient manner.

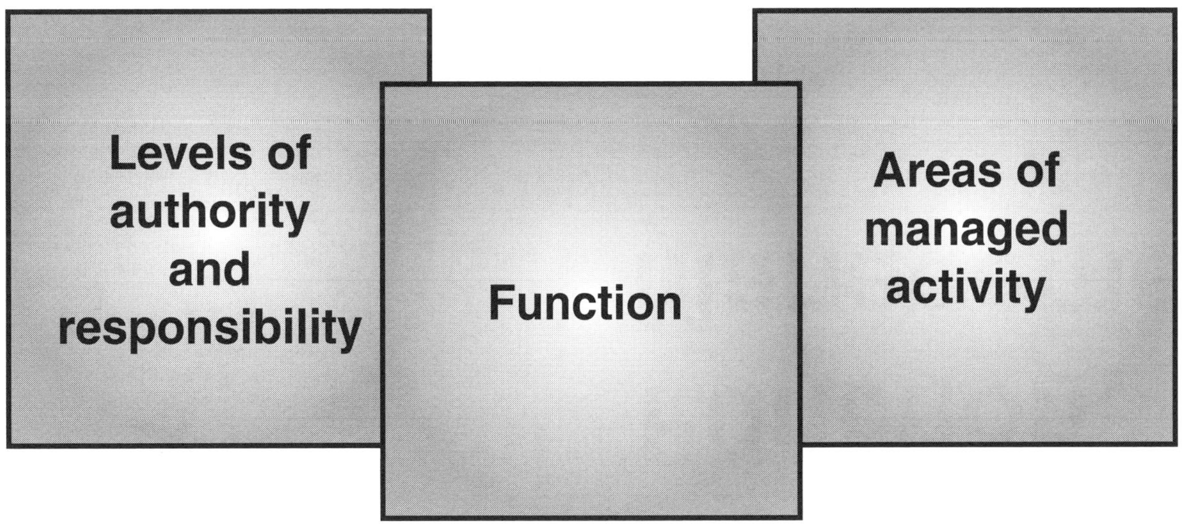

Reproducible Master 8-1

Functions of Management

Planning
setting goals and courses of action

Organizing
structuring and staffing the organization

Actuating
assigning tasks and supervising work

Controlling
measuring results against plans

Levels of Authority and Responsibility

Top management

Middle management

Operating management

Name ______________________________ Date ____________

Period ______________________________ Score ____________

Preparing a Company Organizational Chart

Interview a management employee of a local company about the organization of the company. In the space below, produce an organizational chart.

Reproducible Master 8-4

Name ____________________________ Date___________

Period ___________________________ Score __________

Preparing an Organizational Chart for Your School

Similar to most companies, schools are organized to produce the best results possible and meet goals. There are leaders and managers. They each have responsibilities and people to whom they must report. Interview your teacher. Draw an organizational chart for your school. Place the chart in the space below.

Section 3
Creating Technology

Focus: Presentation of the creation of technology through invention and design, examination of each of the eight steps of the engineering design process.

- ❑ Chapter 9: Invention and Innovation
- ❑ Chapter 10: The Design Process
- ❑ Chapter 11: Identifying Problems
- ❑ Chapter 12: Researching Problems
- ❑ Chapter 13: Creating Solutions
- ❑ Chapter 14: Selecting and Refining Solutions
- ❑ Chapter 15: Modeling Solutions
- ❑ Chapter 16: Testing Solutions
- ❑ Chapter 17: Communicating Solutions
- ❑ Chapter 18: Improving Solutions

Teaching Materials

Text

Chapter 9, pages 204–223
- Chapter Activities, page 218
- Test Your Knowledge, pages 218–219
- Chapter 9 Activity A—Invention and Innovation, pages 220–221

Chapter 10, pages 224–245
- Chapter Activities, page 239
- Test Your Knowledge, page 239
- Chapter 10 Activity A—Design for Space, pages 241–243

Chapter 11, pages 246–259
- Chapter Activities, page 258
- Test Your Knowledge, page 259

Chapter 12, pages 260–275
- Chapter Activities, page 274
- Test Your Knowledge, page 274

Chapter 13, pages 276–299
- Chapter Activities, page 297
- Test Your Knowledge, page 298

Chapter 14, pages 300–317
- Chapter Activities, page 316
- Test Your Knowledge, page 316

Chapter 15, pages 318–345
- Chapter Activities, page 343
- Test Your Knowledge, pages 343–344

Chapter 16, pages 346–365
- Chapter Activities, pages 361–362
- Test Your Knowledge, page 362

Chapter 17, pages 366–393
- Chapter Activities, page 389
- Test Your Knowledge, page 389

Chapter 18, pages 394–409
- Chapter Activities, page 405
- Test Your Knowledge, page 406
- Chapter 18 Activity A—Communication Design, pages 407–409

Tech Lab Workbook

Chapter 9
- Student Activity 9-1—Invention Classification, page 51
- Chapter 9 Study Questions, pages 53–54

Chapter 10
- Student Activity 10-1—Design Careers, pages 55–56
- Chapter 10 Study Questions, page 57

Chapter 11
- Student Activity 11-1—Design Briefs—An Activity in Reverse Engineering, page 59
- Chapter 11 Study Questions, page 61

Chapter 12
- Student Activity 12-1—The Research Process, page 63
- Chapter 12 Study Questions, pages 65–66

Chapter 13
- Student Activity 13-1—Orthographic Sketching, page 67
- Student Activity 13-2—Rendering, page 69
- Student Activity 13-3—Biorelated Feeder, page 71
- Chapter 13 Study Questions, pages 73–74

Chapter 14
- Student Activity 14-1—Elements and Principles of Design, pages 75–76
- Chapter 14 Study Questions, page 77

Chapter 15
- Student Activity 15-1—Classroom Model, page 79
- Chapter 15 Study Questions, page 81

Chapter 16
- Student Activity 16-1—Testing Materials, pages 83–84
- Chapter 16 Study Questions, pages 85–86

Chapter 17
- Student Activity 17-1—Mechanical Drawing, pages 87–90
- Student Activity 17-2—Locker-Organizer Design, page 91
- Chapter 17 Study Questions, page 93

Chapter 18
- Chapter 18 Activity A Worksheet—Communication Design, pages 95–98
- Student Activity 18-1—Product Improvement, page 99
- Chapter 18 Study Questions, page 101

Teacher's Resources

Chapter 9
- Reproducible Master 9-1—Important Inventions in the Twentieth Century, page 121

Chapter 10
- Lesson Slide 10-1—Technological Design (CD only)
- Lesson Slide 10-2—Types of Design (CD only)
- Lesson Slide 10-3—Steps of the Design Process (CD only)

Chapter 11
- Activity Sheet—Identify the Problem, page 111
- Reproducible Master 11-1—The Design Process: Defining the Problem, page 122
- Reproducible Master 11-2—Technological Design Constraints, page 123
- Lesson Slide 11-1—Problems (CD only)
- Lesson Slide 11-2—Constraints (CD only)
- Lesson Slide 11-3—Define the Problem (CD only)

Chapter 12
- Activity Sheet—Research the Problem, page 112
- Reproducible Master 12-1—The Design Process: Gathering Information, page 124
- Lesson Slide 12-1—Gather Information (CD only)

Chapter 13
- Activity Sheet—Create Solutions, page 113
- Reproducible Master 13-1—The Design Process: Developing Possible Solutions, page 125
- Lesson Slide 13-1—Sketching (CD only)
- Lesson Slide 13-2—Rough Sketches (CD only)
- Lesson Slide 13-3—Thumbnails (CD only)
- Lesson Slide 13-4—Pictorial Drawings (CD only)

Chapter 14
- Activity Sheet—Select and Refine the Solutions, page 114
- Reproducible Master 14-1—The Design Process: Selecting Promising Solutions, page 126
- Reproducible Master 14-2—The Design Process: Refining Selected Ideas or Solutions, page 127
- Reproducible Master 14-3—The Design Process: Selecting the Best Solution, page 128
- Lesson Slide 14-1—Select the Best Solution (CD only)
- Lesson Slide 14-2—Rendering (CD only)
- Lesson Slide 14-3—Detailed Sketches (CD only)

Chapter 15
- Alternative Activity—House Design, page 115
- Activity Sheet—Model the Solution, page 116
- Lesson Slide 15-1—Functions of Models (CD only)
- Lesson Slide 15-2—Graphic Models (CD only)
- Lesson Slide 15-3—Mathematical Models (CD only)
- Lesson Slide 15-4—Physical Models (CD only)

Introduction

This section examines the process used to design technological devices and systems. It is essential that the study of technology includes the study of design. The first stage in creating any example of technology involves the engineering design process. All fields of technology have different processes they use to design products, systems, and structures. When they are examined, however, many common elements exist. This book provides an engineering design process that is a combination and revision of many different processes. The design process is a general set of steps that can be used to design any type of product or system. The steps in this process include the following:

1. Identifying problems.
2. Researching problems.
3. Creating solutions.
4. Selecting and refining solutions.
5. Modeling solutions.
6. Testing solutions.
7. Communicating solutions.
8. Improving solutions.

Objectives of the Section

At the end of Section 3, each student should be able to do the following:

- ❑ Understand the term *design*.
- ❑ Summarize the steps in the engineering design process.
- ❑ Recognize a design problem.
- ❑ Create a design brief.
- ❑ Use information to solve a problem.
- ❑ Create rough and refined sketches to develop design solutions.
- ❑ Generate refined design solutions through detailed sketches.
- ❑ Design and build models of design solutions.
- ❑ Use engineering drawings to communicate design solutions.
- ❑ Use appropriate techniques to test design solutions.
- ❑ Apply the process of receiving approval for design solutions.

Suggested Schedule and Length of the Section (2–10 Weeks)

Section 3 introduces the students to the design process. You should have them complete the following assignments:

1. Read Chapters 9–18 in their textbooks and answer associated study questions.
2. Participate in several discussions and presentations about invention and innovation, design, and the eight steps of the engineering design process (identifying, researching, creating, selecting and refining, modeling, testing, communicating, and improving).

3. Complete the scheduled activities and follow-up sheets supporting the concepts of the design process.

Schedule for Section 3—9-Week and 12-Week* Classes	
Day 1	Discussion: "The design process." Assign Chapter 10: "The Design Process" to be read outside of class. Introduce first activity, "Communication Design" or "House Design."
Day 2	Discussion: "Identifying problems." Work on activity. Assign Chapter 13: "Creating Solutions" to be read outside of class.
Day 3	Discussion: "Creating solutions." Work on activity.
Day 4	Work on activity. Assign Chapter 14: "Selecting and Refining Solutions" to be read outside of class.
Day 5	Discussion: "Selecting and refining solutions." Work on activity.
Day 6	Work on activity.
Day 7	Complete and summarize activity. Assign Chapter 15: "Modeling Solutions" to be read outside of class.
Day 8	Discussion: "Modeling solutions." Introduce second activity, "Biorelated Feeder."
Day 9	Work on activity.
Day 10	Complete and summarize activity. Review and summarize section.

* For a 12-week class, use an additional activity and cover three more of the ten chapters. Spread the discussions over the time scheduled for this section.

Schedule for Section 3—18-Week and 36-Week* Classes	
Day 1	Discussion: "Invention and innovation." Assign Chapter 9: "Invention and Innovation" to be read outside of class. Introduce first activity, "Invention and Innovation."
Day 2	Work on activity.
Day 3	Complete and summarize activity. Assign Chapter 10: "The Design Process" to be read outside of class.
Day 4	Discussion: "The design process." Introduce second activity, "Design for Space."
Day 5	Work on activity. Assign Chapter 11: "Identifying Problems" to be read outside of class.
Day 6	Discussion: "Identifying problems." Work on activity.
Day 7	Work on activity. Assign Chapter 12: "Researching Problems" to be read outside of class.
Day 8	Discussion: "Researching problems." Work on activity.
Day 9	Work on activity.
Day 10	Complete and summarize activity. Assign Chapter 13: "Creating Solutions" to be read outside of class.
Day 11	Discussion: "Creating solutions." Introduce third activity, "House Design."
Day 12–13	Work on activity. Assign Chapter 14: "Selecting and Refining Solutions" to be read of outside class.
Day 14	Discussion: "Selecting and refining solutions." Work on activity.
Day 15–17	Work on activity. Assign Chapter 15: "Modeling Solutions" to be read outside of class.
Day 18	Discussion: "Modeling solutions." Work on activity.
Day 19–21	Work on activity. Assign Chapter 16: "Testing Solutions" to be read outside of class.
Day 22	Discussion: "Testing solutions." Work on activity.
Day 23–26	Work on activity. Assign Chapter 17: "Communicating Solutions" to be read outside of class.
Day 27	Discussion: "Communicating solutions." Complete and summarize activity.

Day 28	Assign Chapter 18: "Improving Solutions" to be read outside of class. Introduce fourth activity, "Communication Design."
Day 29	Discussion: "Improving solutions." Work on activity.
Day 30	Complete and summarize activity. Review and summarize section.

*Throughout the class, you might choose, depending on time and resources, to use additional chapter activities to reinforce specific concepts. For a 36-week class, use an additional activity. Spread the discussions over the time scheduled for this section. The sketching handouts can also be used with the appropriate chapters.

Preparing to Teach the Section

Before introducing this section, complete the following tasks:

1. Study this section of the Teacher's Resource.
2. Read Chapters 9–18 in the textbook to gather a complete view of the design process.
3. Prepare the following 10–15 minute discussions and presentations:
   A. Invention and innovation. (Textbook, Chapter 9)
   B. The design process. (Textbook, Chapter 10)
   C. Identifying problems. (Textbook, Chapter 11)
   D. Researching problems. (Textbook, Chapter 12)
   E. Creating solutions. (Textbook, Chapter 13)
   F. Selecting and refining solutions. (Textbook, Chapter 14)
   G. Modeling solutions. (Textbook, Chapter 15)
   H. Testing solutions. (Textbook, Chapter 16)
   I. Communicating solutions. (Textbook, Chapter 17)
   J. Improving solutions. (Textbook, Chapter 18)
4. Select appropriate activities from the following:
   A. Invention and Innovation (Chapter 9), Textbook, pages 220–221.
   B. Invention Classification (Chapter 9), Tech Lab Workbook, page 51.
   C. Design for Space (Chapter 10), Textbook, pages 241–243.
   D. Design Careers (Chapter 10), Tech Lab Workbook, page 55.
   E. Design Briefs—An Activity in Reverse Engineering (Chapter 11), Tech Lab Workbook, page 59.
   F. The Research Process (Chapter 12), Tech Lab Workbook, page 63.
   G. Orthographic Sketching (Chapter 13), Tech Lab Workbook, page 67.
   H. Rendering (Chapter 13), Tech Lab Workbook, page 69.
   I. Biorelated Feeder (Chapter 13), Tech Lab Workbook, page 71.
   J. Elements and Principles of Design (Chapter 14), Tech Lab Workbook, page 75.
   K. Classroom Model (Chapter 15), Tech Lab Workbook, page 79.
   L. House Design (Chapter 15), Teacher's Resource, page 115.
   M. Google SketchUp® Software (Chapter 15), External Resource.
   N. Testing Materials (Chapter 16), Tech Lab Workbook, page 83.
   O. Mechanical Drawing (Chapter 17), Tech Lab Workbook, page 87.
   P. Locker-Organizer Design (Chapter 17), Tech Lab Workbook, page 91.
   Q. Communication Design (Chapter 18), Textbook, pages 407–409.
   R. Product Improvement (Chapter 18), Tech Lab Workbook, page 99
5. Prepare materials and supplies for selected activities and presentations.
6. Work through the activities and student study questions in the textbook, Tech Lab Workbook, and Teacher's Resource.

Presenting the Section

This section will be taught differently, depending on the length of the course. In a 9-week course, there are 2 weeks for this section. In 2 weeks, the students have time to go through a small design project. There is not time for an in-depth look at any of the skills needed in the design industry. This issue is also a problem in the 36-week course, in which 10 weeks can be spent on this section. In this amount of time, it might seem important to delve deeper into concepts such as sketching, rendering, and technical drawings. This course, however, is *not* a course in technical graphics. The development of drawing skills should not be overemphasized. The overall understanding of the process of design is more important than the students' drawing and drafting abilities. The 36-week course simply offers the opportunity to conduct several design experiences. To remain on schedule throughout this section, do the following:

- ❑ Keep all presentations between 10 and 15 minutes.
- ❑ Demonstrate only fundamental drawing skills.
- ❑ Keep design problems simple. Complex problems only increase the amount of time it takes students to complete them.
- ❑ Prepare all materials and supplies in advance to ensure that the students have everything they need to complete the activities.
- ❑ Maintain the focus on the design process.

Adjustments for Students with Special Needs and Gifted Students

The activities in this section are both directive and open-ended. Each of the activities works well as either individual or cooperative group work. Using group work is an effective way to adjust instruction for students of varying abilities and interests. Students can be assigned so a mix of abilities, learning styles, or motivation levels is achieved. A typical group can include a gifted student, an average student, and a student with special needs.

Give special attention to students who might require more structure, especially during open-ended design projects. It is important, however, that the activities are free enough that the design solutions are different. Design, by nature, must be allowed to be creative and unique.

Evaluation Procedures

This section stresses the understanding of the design process. The focus should be on the understanding of the process used to design solutions. There are no right or wrong solutions. Some solutions, however, fit the design brief better than others. This makes evaluation of this section subjective. Therefore, students should be evaluated on the following factors:

- ❑ Contributions to group activities. This can be done with daily logs (diaries) the students maintain, teacher observations, and student feedback.
- ❑ Creativity and divergent thinking shown during the design activities.
- ❑ Laboratory sheets, sketches, and designs completed during the activities.
- ❑ Participation in discussions conducted during the activities.
- ❑ Scores on quizzes and chapter questions.

Implementing the Laboratory Activities

Several different activities have been developed for this section. Each chapter has at least one student activity, found in the Tech Lab Workbook. This activity focuses on the content specific to that chapter. These activities explore specific skills related to the design process (such as researching, sketching, modeling, and testing). There are also additional activities, found at the end of Section 3 in the textbook, in the Tech Lab Workbook, and in this Teacher's Resource, that focus on the entire design process. These are best completed by the students after all the design chapters have been studied. On the other hand, each step of the design process can be completed after the students study the corresponding chapter. The design problems cover many different contexts. Some of the problems have been developed as structured activities. The remaining activities are open-ended design briefs. There is not time in your course to utilize all the chapter activities. So, when selecting activities, try to choose a sampling of structured and open-ended activities to make sure all students are reached. There are also a few short classroom handouts included that can be used for the review of concepts, if time permits.

There are several design-related activities. Some of them are larger projects that include all aspects of the design process detailed in Chapters 11–18 of the textbook. The activities are designed so the design-process activity sheets, in the Teacher's Resource, can be used for each stage of the design process. The students can complete all the stages of the design activity after covering all the chapters. If you would rather, each step can be presented to the students at the conclusion of each design chapter. Some of the activities are smaller activities highlighting only several steps of the design process. These are best used in shorter courses in which there is not the time to cover all the design chapters and for students to complete activities related to each step.

While completing the activities, the students should be encouraged to think "outside the box" and develop new and creative solutions. This does not mean all solutions must be completely original and never seen before. The students' solutions can be innovations or adaptations of existing products and ideas.

The activities are intended to reinforce the chapter content by allowing for practice or having the students explore the content in greater depth. For most chapters, the activities are more than just worksheets that can be completed in an evening. Several of the activities are research projects requiring time and resources. These and other activities in this Teacher's Resource also require teacher direction and potentially additional instruction.

Procedures for Directing Each Suggested Activity

Chapter 9: Invention and Innovation, Textbook, pages 220–221

This activity is a review of the history of inventions. The review should be used in the longer courses that cover Chapter 9, "Invention and Innovation." This activity asks the students to research an invention and explain, not only the invention's history, but also the inventions leading up to it. They will also describe any inventions that have developed from their invention, as well as the impacts of the invention. In order for the students to complete this activity, they must have access to research materials. The Internet and basic invention books, as well as encyclopedias, are all helpful to the students.

Chapter 9: Invention Classification, Tech Lab Workbook, page 51

"Invention Classification" is a review of the types of invention. This activity asks the students to list and locate images of different types of inventions. The review can be done either in or out of class and in groups or individually. Image sources such as magazines, encyclopedias, and the Internet are helpful to students. As the teacher, you need to select the presentation style of this assignment. Inform the students how to create their presentation (such as with the PowerPoint® presentation graphics program, on poster board, or as a digital story).

Chapter 10: Design for Space, Textbook, pages 241–243

This is an open-ended design activity that challenges the students to develop a model of a space station with living quarters, recreational areas, production facilities, office and managerial areas, mechanical support facilities, and a flight command center. To complete the activity, allow the students to work in groups of two to three students. Give each group a No. 10 tin can (a large can from the school cafeteria or a local restaurant) and a supply of poster board, construction paper, hollow tubes or PVC pipes, and dowels. Have each group then do the following:

1. Read the design brief in their textbooks.
2. Use the design process to design a space station.
3. Prepare a model of the station.
4. Present the model to the class.

Chapter 10: Design Careers, Tech Lab Workbook, pages 55–56

In this activity, students explore design careers. Prior to conducting the activity, engage your students in a discussion or brainstorming session on the different types of designers. This will ensure that students research a range of careers. You might also want to limit the number of students that can research each specific career to make sure there is variety in the presentations. The students will need library or Internet access to conduct the research. They will also need guidance on how they are to present their findings.

Chapter 11: Design Briefs—An Activity in Reverse Engineering, Tech Lab Workbook, page 59

Students are to create a design brief for an existing product in this activity. In the introduction to this activity, you might want to go through the activity as a class so students understand the process. This activity requires creative thinking and is a good activity to conduct in small groups. To extend the activity, you can have the students examine competing products and determine if similar design briefs could have been used to design both products.

Chapter 12: The Research Process, Tech Lab Workbook, page 63

In this activity, students design and conduct a study. Several examples of potential studies are listed in the procedure section of the activity. As the teacher, you might want to walk the students through one of those examples (or an example of your own). It might also be helpful if you create an example set of data and teach the students how to create a chart. This should be conducted as a group project.

Chapter 13: Orthographic Sketching, Tech Lab Workbook, page 67

This activity sheet is an activity intended to give students practice at creating orthographic sketches. Colored pencils are the only resources needed.

Chapter 13: Rendering, Tech Lab Workbook, page 69

"Rendering" is an activity intended to give students practice at creating renderings. Colored pencils are the only resources needed.

Chapter 13: Biorelated Feeder, Tech Lab Workbook, page 71

This is an open-ended design brief. The activity allows students the freedom to create any type of automatic feeder they desire. They can even choose to design an improved model of a product they use to feed a cat or dog at home. Students should be encouraged to use their creativity and imagination. The design sheets in the Teacher's Resource can be used to guide the students' work. In this activity, the students will do the following:

1. Select the types of devices they are designing.
2. Create sketches of possible designs.
3. Create displays of their final designs.
4. If time permits, you can choose to have the students build models of their designs.

Chapter 14: Elements and Principles of Design, Tech Lab Workbook, pages 75–76

In this activity, the students will find examples of the elements and principles of design. The students will define and locate images illustrating the elements or principles of design. This activity can be done either in or out of class and in groups or individually. Image sources such as magazines, encyclopedias, and the Internet are helpful to students. As the teacher, you need to inform students how to create their presentation (such as with the PowerPoint® presentation graphics program, on poster board, or as a digital story).

Chapter 15: Classroom Model, Tech Lab Workbook, page 79

In this activity, the students will create a physical model of the classroom. Students need tape measures, modeling materials (such as chipboard and matboard), scales, straight edges, and hobby knives. They might also need additional instruction in using scales. To extend this activity, you might want to assign a specific scale for all students to use and then assign different classrooms to various groups of students. When completed, the students' models should fit together to create a representation of a wing of the school.

Chapter 15: House Design, Teacher's Resource, page 115

This is a structured design activity that allows for different solutions. The activity focuses mainly on the design and modeling stages of the design process. Each design step, however, is touched on. The students will design a floor plan for a single-story home under 1800 square feet. The size restriction is large enough to allow the students to add rooms they would enjoy, such as game rooms or exercise rooms. The size is small enough, however, to keep the designs and models at a reasonable size. This Teacher's Resource includes a design brief on this activity. The brief also includes a client sheet. This sheet will help the students define the problem, criteria, and constraints of the house. The teacher should lead the students through the process below, making sure the materials are prepared in advance. The students need the following supplies: paper; 1/4″ graph paper; masking or Scotch® tape; rubber cement; corrugated cardboard or foam-core board; hobby or utility knives; white, wood, or hot glue; and paint.

Procedure

Identifying a problem

1. Have the students review the client sheet. This sheet should help them determine the needs (design problem) of the homeowner.

Researching the solution

2. Distribute sample one-story house plans to the students. The plans can be collected from house-plan books or local home builders.
3. Discuss simple room-layout concepts, such as quiet zones and loud zones, service and utility areas, the use of storage space, and typical room sizes.

Creating solutions

4. Have each student sketch a floor plan, on plain paper, that fulfills the requirements of the homeowner. At this point, the students should not worry too much about the scale of the drawing. They should just get their basic ideas on paper.

Selecting and refining solutions

5. At this point, you should make the decision to allow the students to work in groups or continue individually. If they are working in groups, the students should gather their designs and select the most promising design.
6. Distribute 1/4″ graph paper. The students will need several sheets for each house.
7. Have the students draw their houses at a scale of 1′ for every one box on the graph paper.
8. Watch the designs they are creating to make sure the room sizes are appropriate.
9. Have the students mark the doors and windows and label the names of each room.

Modeling the solution

10. Distribute sheets of cardboard. Local container companies are a good source of cardboard. Thin, corrugated cardboard works the best because these models are small (1/4″ = 1′).
11. Have the students use rubber cement to glue their plans to the cardboard and cut around the exteriors of their homes.
12. Prepare strips of cardboard that are 2″ tall and varied lengths for the students to use as walls.
13. The students should cut their walls to length and then cut out the doors and windows. Doors should be cut 3/4″ wide and about 1 5/8″ tall. Windows can be cut as needed.
14. Walls should then be painted. If time permits, have the students decorate the walls of each room differently, as opposed to painting every wall the same color.

15. After the walls have dried, the students should assemble the walls using white, wood, or hot glue.
16. If time permits, have the students build and paint a roof that lifts off the house.

Testing the solution

17. Have the students examine their houses to make sure the houses meet the needs of the client.

Communicate the solution

18. The students should then draw front and side views of their houses, as the houses would be seen finished from the road. They can add trees and landscaping to the drawings.
19. Have the students present their homes to the rest of the class.

Improving the solution

20. After viewing the presentations, ask the students how they might have improved their designs.

Chapter 15: Google SketchUp Software, External Resource

Your students can create computer models using Google SketchUp software (http://sketchup.google.com), your students can create computer models. Google SketchUp software is a surface-modeling program freely available to you and your students. There is also a community of users who have created libraries of tutorials and example models that will be beneficial to your students. This external resource can be extremely beneficial to your courses, if you are interested in exploring computer models with your students.

Chapter 16: Testing Materials, Tech Lab Workbook, pages 83–84

Students will be active in the material testing of several material properties. This is a good small-group activity. Each group needs a 1/2″ × 4″ × 12″ piece of foam, plywood, hardwood, and plastic. (The size can be different. All materials, however, should be of a similar size.) To conduct the tests, the students need a hammer, quarter, clamp, and weight. You might choose to set up stations in which the students rotate through each station and test their materials. To extend this activity, you can have your students brainstorm other material properties that can be tested and ways in which they would perform those tests.

Chapter 17: Mechanical Drawing, Tech Lab Workbook, pages 87–90

In this activity, the students will create mechanical drawings. Students need mechanical drawing tools (a drawing board, a T square, triangles, and a compass) or a computer-aided design (CAD) software program. They might also need additional instruction in using the mechanical drawing tools. To extend this activity, you can give students physical objects to measure and draw.

Chapter 17: Locker-Organizer Design, Tech Lab Workbook, page 91

This is an open-ended design project that challenges the students to design an organizer for their lockers. The students should complete each of the design steps in the creation of their solution. This Teacher's Resource has a page that can be used for each step of the design process. These generic design sheets can be used for this activity or any other design project.

Chapter 18: Communication Design, Textbook, pages 407–409, and Tech Lab Workbook, pages 95–98

This is a prescriptive activity that challenges the students to design a television commercial to promote your class. The activity is organized around the design process outlined in Section 3. At several steps, the students are required to make design decisions. It is important for the students to base their decisions on the design problem and design brief, instead of simply selecting the design they like the most. This activity uses both group and individual work. Whenever the students are designing the storyboards, or possible solutions (in another activity), they should be encouraged to work individually. This helps the students gain a better understanding of the design process.

Chapter 18: Product Improvement, Tech Lab Workbook, page 99

This is an activity in which students brainstorm potential product improvements. Once the students have developed a list of improvements, they are to create a package highlighting the improvements. You might have the students collect different product packages and examine them as a class, prior to the design of their own packages. The students need materials to create the packages, such as poster board, cardboard, markers, colored pencils, tape, and glue.

Answer Key for Section 3

Chapter 9—Invention and Innovation

Test Your Knowledge, Textbook, pages 218–219

1. An invention is a new and unique product.
2. The Industrial Revolution was important because many of the inventions of that time have had lasting effects on the world.
3. Discoveries are made when something that occurs naturally is first noticed. Inventions are designed and created by humans.
4. devices and machines, products, and processes
5. False
6. A problem is anything that can be made better through change.
7. False
8. True
9. Answers will vary. Some acceptable answers are curious, creative, imaginative, and knowledgeable.
10. social
11. A financial invention
12. leisure
13. True
14. innovation
15. D. An invention used for something other than the purpose for which it was intended.

Study Questions, Tech Lab Workbook, pages 53–54

1. invention
2. Industrial Revolution
3. B. Gravity.
4. devices and machines, products, and processes
5. problem-solving process
6. Identify the problem, collect information, think creatively, experiment, and review
7. Answers will vary. Some acceptable answers are curious, creative, imaginative, and knowledgeable.
8. social invention
9. leisure invention
10. Each team member has different abilities and knowledge.
11. C. Invention.
12. A. Adaptation.
13. B. Innovation.

Chapter 10—The Design Process

Test Your Knowledge, Textbook, page 239

1. Answers will vary. Some acceptable answers are a desk, the carpet, the lights, clothes, and a painting.
2. purpose
3. False
4. Artistic design is the type of design used to solve problems that the artist chooses.
5. Engineering
6. form, function
7. Answers will vary. Some acceptable answers are fashion designers, graphic designers, and chemists.
8. False
9. D. None. This process never ends.
10. Identifying the problem (writing the need or want as a problem statement), researching the problem (learning about the need or want), creating solutions (developing ideas), selecting and refining solutions (listing the aspects of design and creating a design), modeling the solution (creating tables, formulas, charts, and mock-ups), testing the solution (determining whether or not the design works), communicating the solution (developing

a bill of materials, specification sheets, and drawings), and improving the solution (modifying the design)

Study Questions, Tech Lab Workbook, page 57

1. Design
2. Answers will vary. Some acceptable answers are desks, carpet, lights, clothes, and paintings.
3. False
4. C. Content.
5. A. Form.
6. B. Function.
7. system
8. product
9. Identifying the problem, researching the problem, creating solutions to the problem, selecting and refining solutions, modeling the solution, testing the solution, communicating the solution, and improving the solution

Chapter 11—Identifying Problems

Test Your Knowledge, Textbook, page 259

1. True
2. Answers will vary. Some acceptable answers are to talk to someone on the phone, to travel to the Moon, to receive a new skateboard, and to sit in a softer chair.
3. problem
4. B. opportunity.
5. Social problems (those dealing with society), personal problems (those affecting individuals), and technological problems (those that can affect individuals and groups of people)
6. False
7. True
8. To list the requirements of the solution
9. Technical constraints (the size and shape of the object and the materials that will be used), financial constraints (the costs of the design), legal constraints (compliance with laws and regulations), time constraints (the amount of time the design process can take), and production constraints (how the device can be built)
10. True

Study Questions, Tech Lab Workbook, page 61

1. need
2. B. Want.
3. A. Need.
4. A. Need.
5. B. Want.
6. problem, opportunity
7. social
8. personal
9. False
10. False
11. criteria
12. Technical, financial, legal, time, and production constraints
13. design brief

Chapter 12—Researching Problems

Test Your Knowledge, Textbook, page 274

1. False
2. Library
3. Survey
4. False
5. In the design brief
6. data
7. information
8. True
9. conclusions
10. Listing the problem involves identifying and understanding the problem statement. Gathering data can be done through library, survey, or experimental research. Analyzing the data includes reviewing and organizing the data gathered. Making a conclusion includes interpreting, evaluating, and determining the accuracy of the information and then making a list of the findings.

Study Questions, Tech Lab Workbook, pages 65–66

1. research
2. Library, survey, and experimental research
3. market
4. scientific method
5. True
6. data

7. C. look
8. information
9. A. Line graph.
10. C. Bar graph.
11. B. Pie graph.
12. conclusions

Chapter 13—Creating Solutions

Test Your Knowledge, Textbook, page 298

1. ideation
2. graphic organizer
3. False
4. True
5. Any two of the following: isometric, oblique, or perspective
6. A. Evaluate individually.
   B. Evaluate individually.
7. shading
8. Evaluate individually.
9. Perspective sketches show how objects look to the eye.

Study Questions, Tech Lab Workbook, pages 73–74

1. Ideation
2. A. Brainstorming
3. C. Using a graphic organizer.
4. B. Questioning.
5. rough sketch
6. refined sketches
7. Visualizing the object, blocking out shapes, adding an outline, and drawing design features.
8. three
9. Box, cylinder, sphere, cone, and pyramid
10. shading
11. isometric
12. isometric
13. perspective

Chapter 14—Selecting and Refining Solutions

Test Your Knowledge, Textbook, page 316

1. False
2. Appearance, function, human factors engineering, production, safety, environmental impact, and finances
3. ergonomics
4. To show the refined sketch of the design and list the aspects of design
5. evaluation grid
6. shape
7. True
8. The use of two opposite colors to bring attention to something.—B. Contrast.
   The effect of motion.—D. Rhythm.
   The weight of elements on each side of a design.—A. Balance.
   The blending of a design.—C. Unity.
9. False
10. Size, location, and shape information

Study Questions, Tech Lab Workbook, page 77

1. external selection
2. aspects
3. C. Appearance.
4. E. Function.
5. B. Human factors.
6. A. Production.
7. D. Finances.
8. True
9. evaluation grids
10. value
11. texture
12. balance
13. rendering
14. Size, location, and shape

Chapter 15—Modeling Solutions

Test Your Knowledge, Textbook, pages 343–344

1. product, system, process, idea
2. communicate, experiment, organize
3. graphic model
4. mathematical
5. A. Wireframe models use lines to represent the edges of an object, while remaining see-through.
   B. Surface models are wireframe models with coverings, though they are still hollow.
   C. Solid models have edges, surfaces, and interior features and can be used to represent the insides and outsides of designs.

6. physical models
7. False
8. A material made by gluing thin sheets of wood together.—A. Plywood.
   A single thin sheet of wood.—B. Veneer.
   A material made with paper on the outside and foam on the inside.—D. Foam-core board.
   A material also known as *museum board*.—C. Matboard.
   A plastic material too thick to cut with a knife.—E. Acrylic.
   Two different types include modeling and styling.—F. Clay.
   Examples will vary. Some acceptable answers are as a base (plywood), glued on top of the base (veneer), for a final mock-up (foam-core board), to add features to the model (matboard), to represent glass (acrylic), and to create a complete model (clay).
9. False
10. 1. Determine the type of model and material.
    2. Rough out the shape.
    3. Smooth the shape.
    4. Assemble the model.
    5. Finish the model.

Study Questions, Tech Lab Workbook, page 81

1. model
2. Communication, experimentation, and organization
3. D. Graphic.
4. A. Mathematical.
5. C. Computer.
6. B. Physical.
7. prototype
8. Wood, paper, plastic, and metal sheet materials
9. foam
10. True
11. Determine the type of model and material, rough out the shape, assemble the model, and finish the model.
12. Rapid prototyping

Chapter 16—Testing Solutions

Test Your Knowledge, Textbook, page 362

1. Testing a solution is necessary because the solution must solve a problem.
2. False
3. After reviewing the results of a test
4. True
5. B. Performance is how well the solution works.
6. Function describes how the solution works. Human engineering examines the way people interact with the solution. Economics check to make sure the designers will make money. Appearance is how the product looks. Safety ensures that the solution is safe for consumers, workers, and the environment.
7. Laboratory, field, and computer
8. True
9. Determine the purpose, choose the test, collect materials and equipment, conduct the test and gather data, analyze the data, and evaluate the results
10. True

Study Questions, Tech Lab Workbook, pages 85–86

1. True
2. test
3. assessment
4. True
5. Function, human engineering, economics, appearance, and safety
6. durability
7. material properties
8. A. laboratory
9. B. field test
10. valid
11. Determine the purpose, choose the test, collect materials and equipment, conduct the test and gather data, analyze the data, and evaluate the results.
12. Examples can vary. An acceptable answer is the American Society for Testing and Materials (ASTM).
13. assessment
14. trade-off

Chapter 17—Communicating Solutions

Test Your Knowledge, Textbook, page 389

1. False
2. True
3. Answers will vary. Some acceptable answers are toasters, space shuttles, and automobiles, for engineering drawings, and floor plans, building details, and bridges, for architectural drawings.
4. Detailed, assembly, and schematic
5. True
6. multiview drawings
7. Answers will vary. Some acceptable answers are drawing boards, T squares, 45° triangles, 30°–60° triangles, compasses, dividers, templates, scales, erasers, and pencils.
8. False
9. True
10. The designers must present a written report highlighting the features of the design to management. This report includes the expectations and costs for the design, along with final drawings of the solution. The designers distribute this report and give an oral presentation on the solution.

Study Questions, Tech Lab Workbook, page 93

1. bill of materials
2. Mechanical
3. specification sheet
4. Architectural
5. A. Detailed.
6. C. Assembly.
7. B. Schematic.
8. Orthographic
9. alphabet of lines
10. False
11. cabinet oblique drawing
12. Answers will vary. Some acceptable answers are T squares, triangles, and compasses.
13. True
14. True

Chapter 18—Improving Solutions

Test Your Knowledge, Textbook, page 406

1. False
2. Answers will vary. Some acceptable answers are antique cars, outdated computers, and old wiring.
3. Profit
4. Competition is a situation in which two or more companies sell the same types of products.
5. False
6. negative
7. True
8. more than one
9. True
10. A designer might have to start over because the product does not fit the users' needs or function correctly.

Study Questions, Tech Lab Workbook, page 101

1. False
2. profit
3. patch
4. Answers will vary. An acceptable answer is food.
5. Consumers, competitors, government organizations, and internal sources
6. standardization
7. False

Captions for Reproducible Masters

- ❑ **Reproducible Master 9-1.** Important Inventions in the Twentieth Century—Many inventions in the past century have impacted the way we live today.
- ❑ **Reproducible Master 11-1.** The Design Process: Defining the Problem—Some main points to remember when defining the problem are listed.
- ❑ **Reproducible Master 11-2.** Technological Design Constraints—There are several types of constraints to consider in any design.
- ❑ **Reproducible Master 12-1.** The Design Process: Gathering Information—Some

important steps are listed for gathering information.

- ❑ **Reproducible Master 13-1.** The Design Process: Developing Possible Solutions—Some main points to remember when developing possible solutions are listed.
- ❑ **Reproducible Master 14-1.** The Design Process: Selecting Promising Solutions—Some main points to remember when selecting promising solutions are listed.
- ❑ **Reproducible Master 14-2.** The Design Process: Refining Selected Ideas or Solutions—Some main points to remember when refining selected solutions are listed.
- ❑ **Reproducible Master 14-3.** The Design Process: Selecting the Best Solution—Some main points to remember when selecting the best solution are listed.
- ❑ **Reproducible Master 16-1.** The Design Process: Testing Design Solutions—Some main points to remember when testing design solutions are listed.
- ❑ **Reproducible Master 17-1.** The Design Process: Interpreting the Solution—Some main points to remember when interpreting the solution are listed.
- ❑ **Reproducible Master 17-2.** Making an Oblique Drawing—The steps for making an oblique drawing are shown.
- ❑ **Reproducible Master 17-3.** The Design Process: Presenting Solutions for Approval—Some main points to remember when presenting solutions are listed.

Captions for Lesson Slides (CD Only)

- ❑ **Lesson Slide 10-1.** Technological Design—We live in a designed world. The artifacts we design are used in systems we create. These designed systems are then placed in environments we have designed to solve a specific problem.
- ❑ **Lesson Slide 10-2.** Types of Design—Artistic design and engineering design are the two main types of design.
- ❑ **Lesson Slide 10-3.** Steps of the Design Process—The design process uses a series of steps to design a new product or system.
- ❑ **Lesson Slide 11-1.** Problems—A problem is a situation that can be made better through an improvement.
- ❑ **Lesson Slide 11-2.** Constraints—The five main types of constraints are shown.
- ❑ **Lesson Slide 11-3.** Define the Problem—There are several steps involved in defining the problem.
- ❑ **Lesson Slide 12-1.** Gather Information—Researching can be done in a few different ways. The three major types of research are listed.
- ❑ **Lesson Slide 13-1.** Sketching—The main functions of sketching are listed.
- ❑ **Lesson Slide 13-2.** Rough Sketches—The main features of rough sketches are listed.
- ❑ **Lesson Slide 13-3.** Thumbnails—The main features of thumbnail sketches are listed.
- ❑ **Lesson Slide 13-4.** Pictorial Drawings—Drawings in three dimensions are pictorial drawings. The three major types are oblique, isometric, and perspective.
- ❑ **Lesson Slide 14-1.** Select the Best Solution—The best design should be chosen by looking at the advantages and disadvantages of each one. The main categories of criteria are listed.
- ❑ **Lesson Slide 14-2.** Rendering—A rendering shows the overall appearance of the solution. The main features of rendering are shown.
- ❑ **Lesson Slide 14-3.** Detailed Sketches—Detailed sketches show the dimensions of the solution. The major functions of this type of sketch are listed.
- ❑ **Lesson Slide 15-1.** Functions of Models—Models are used to communicate, experiment, and organize.
- ❑ **Lesson Slide 15-2.** Graphic Models—Graphic models are visual representations on paper.

- ❑ **Lesson Slide 15-3.** Mathematical Models—Mathematical models use symbols, numbers, and letters to represent values in a formula.
- ❑ **Lesson Slide 15-4.** Physical Models—Physical models are actual, three-dimensional replicas of a design. Mock-ups and prototypes are the two main types of physical models.
- ❑ **Lesson Slide 16-1.** Design Assessment—Assessments help designers measure something about their solution.
- ❑ **Lesson Slide 17-1.** Pictorial Drawings—Pictorial drawings include isometric, oblique, and perspective drawings.
- ❑ **Lesson Slide 17-2.** Isometric Drawing—The steps of isometric drawing are illustrated.
- ❑ **Lesson Slide 17-3.** Oblique Drawing—The steps of oblique drawing are illustrated.
- ❑ **Lesson Slide 17-4.** One-Point Perspective—The steps for drawing a one-point perspective drawing are illustrated.
- ❑ **Lesson Slide 17-5.** Two-Point Perspective—The steps for drawing a two-point perspective drawing are illustrated.

Name ______________________ Date __________

Period ______________________ Score __________

Chapter 11 Activity Sheet

Identify the Problem

Statement of Problem

Criteria

Constraints

Name ______________________ Date __________

Period ______________________ Score __________

Chapter 12 Activity Sheet

Research the Problem

List the information found in the following locations:

- ❑ Library

- ❑ Internet

- ❑ Interviews

- ❑ Other: ______________________

Name ______________________ Date __________

Period ______________________ Score __________

Chapter 13 Activity Sheet

Create Solutions

Sketch three different possible designs below.

Design 1

Design 2

Design 3

Name_________________________ Date __________

Period _______________________ Score__________

CHAPTER 14 ACTIVITY SHEET

SELECT AND REFINE THE SOLUTIONS

Selecting the Best Solution

In the boxes below, list the positive and negative aspects of each possible design.

Design 1	Design 2	Design 3
Positive	Positive	Positive
Negative	Negative	Negative

Refining the Solution

In the space provided, draw a refined sketch of the selected design.

Name ________________________ Date __________

Period ________________________ Score __________

Chapter 15 Alternative Activity

House Design

Design Problem: You are to design and model a house fitting the needs of the following client.

Client Name: Robert Zimmerman

House Type: One story

Area: Maximum—1800 square feet for living area, 2200 square feet with garage

Rooms Needed:

- ❑ Master bedroom
- ❑ Master bathroom
- ❑ Two additional bedrooms
- ❑ One additional bathroom
- ❑ Great room or living room
- ❑ Dining room or dining area
- ❑ Kitchen
- ❑ Garage
- ❑ Closets
- ❑ Laundry room

Optional rooms:

- ❑ Home office
- ❑ Den
- ❑ Game room
- ❑ Library
- ❑ Pantry
- ❑ Fitness room
- ❑ Porch or deck

Garage: Two car

Name________________________ Date __________

Period ________________________ Score__________

Chapter 15 Activity Sheet

Model the Solution

List the materials needed to create a model of the design:

Collect the materials. Build the model.

Name ______________________ Date __________

Period ______________________ Score __________

Chapter 16 Activity Sheet

Test the Solution

Describe the test you will conduct on the solution.

Answer the following questions about the results of the test:

1. Does the product function correctly?

   ____

2. Is the product safe?

   ____

3. Does the product pass the test? Explain.

   ____

   ____

   ____

   ____

Name____________________________ Date ___________

Period ___________________________ Score___________

Chapter 17 Activity Sheet

Communicate the Solution

Draw a dimensioned sketch or drawing in the space below.

Name ____________________ Date __________

Period ____________________ Score __________

Chapter 18 Activity Sheet

Improve the Solution

Answer the questions below.

1. How well does the solution fit the design brief?

____

____

____

____

2. What changes can be made to make your solution better?

____

____

____

____

Draw the improved final solution. Label the features you have changed.

Important Inventions in the Twentieth Century

Date	Development	Developer	Impact
1902	Air conditioning	Willis Carrier	Made dwellings and automobiles more comfortable
1903	Functioning airplane	Orville and Wilbur Wright	Made travel more efficient and rapid
1904	Vacuum tube	John Flemming	Changed alternating current to direct current; opened the way for the development of the radio and other types of electronic devices
1908	Affordable car	Henry Ford	Made it possible for many people to own cars
1909	Successful plastic	Leo Baekeland	Bakelite® plastic became the first of a long list of plastics that are light and strong, serve as a substitute for many metals, and are not affected by corrosion or rust
1912	Heterodyne circuit	Reginald A. Fessenden	Important improvement in radio reception
1924	Iconoscope	Vladimir K. Zworykin	Made modern television possible
1928	Penicillin	Alexander Fleming	Improved life for everyone by providing a powerful antibiotic to fight infections
1930	Jet engine	Frank Whittle	Provided more powerful propulsion system for aircraft
1931	Electron telescope	Ernst Ruska	Improved study of astronomy by providing greater magnification of distant stars and planets
1934	Nylon	Wallace Hume Carothers	Provided the modern wear-resistant material used in manufacturing many products, including clothing and automobile parts
1935	Radar	Robert A. Watson-Watts	Made air travel safer
1937	Xerography	Chester Carlson	First method of photocopying; has led to the development of the fax machine
1939	Mass-produced helicopter	Igor Sikorsky	First aircraft, beside balloons, that could hover in the air; could take off from and land on a small landing pad; made short flights feasible
1940	Color television	Peter Carl Goldmark	Improved reception of mass communications and entertainment
1941	Dacron® polyester	John Rex Whinfield	Became popular fabric for clothing
1942	Nuclear reactor	Enrico Fermi	Provides important percentage of electrical power worldwide
1947	Holography	Dennis Gabor	Three-dimensional photography
1948	Transistor	William Shockley, Walter Brattain, and John Bardeen	Replaced vacuum tubes in radio and television; made space communications possible; led to smaller, more compact computers
	33 1/3–revolutions per minute (rpm) phonograph records	Peter Carl Goldmark	More music could be recorded in less space
1957	Laser	Gordon Gould	Important development in all technologies, including communications, machining technology, and construction

Reproducible Master 9-1

The Design Process: Defining the Problem

- ❑ Define the problem. (Example: Your books do not stand upright.)
- ❑ The problem to be solved must be stated in such general terms that the designer is free to explore all design possibilities.
- ❑ The problem statement should clearly say what is needed.
- ❑ Examples:

  Restrictive—"Design a set of bookends."

  Nonrestrictive—"Design a device to hold books upright."

Technological Design Constraints

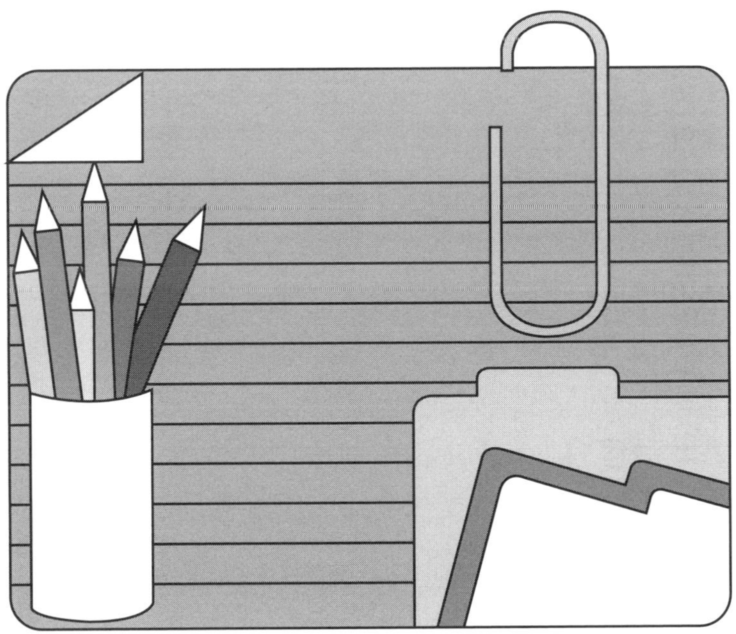

- ✔ Technical constraints
- ✔ Production constraints
- ✔ Market constraints
- ✔ Financial constraints
- ✔ Environmental constraints
- ✔ Time constraints

Reproducible Master 11-2

The Design Process: Gathering Information

- ❑ What information is needed? How can you gather it?
- ❑ Decide what questions need answers.
- ❑ Make a list of literature resources. Include the following sources:
  - ❑ Books.
  - ❑ Magazines.
  - ❑ Clipping services.
  - ❑ Company records.
  - ❑ Study designs of others. Also, study designs from catalogs and books.
- ❑ Seek opinions of others.
  - ❑ Interview possible users.
  - ❑ Develop a questionnaire.

The Design Process: Developing Possible Solutions

- ❑ The designer reviews facts from the information collected.
- ❑ Generally, the designer starts by sketching ideas on paper.
- ❑ Good designers try to capture as many ideas as possible, sketching them rapidly as the ideas occur.
- ❑ Rough sketches merely show how the product will look in a pictorial view.

Reproducible Master 13-1

The Design Process: Selecting Promising Solutions

- This is one of the points in the design process in which someone must make decisions. These decisions are which sketches to approve.
- The most promising designs are selected for further development.
- These decisions must be made carefully, as wrong choices are costly.
- No one wants to waste resources on products that are not needed or wanted.
- Usually, several managers and product experts review the designs.
- At this stage, design improvements are discussed and developed.
- Sometimes, groups of people are asked to brainstorm to resolve a design problem.
- Each suggestion for improvement is carefully reviewed.
- Some ideas can be used to further improve the design.

The Design Process: Refining Selected Ideas or Solutions

- ❑ The artist, designer, or drafter resketches the most promising ideas.
- ❑ Refined sketches clarify and extend the ideas suggested in the rough sketches.
- ❑ The refinement might center on one rough sketch. The modification might, however, combine ideas from several sketches.
- ❑ The purpose of the refined sketches is to narrow down and improve the possible solutions.
- ❑ As with the rough sketches, the refined sketches are generally pictorial.
- ❑ The drawings can be oblique or isometric.
- ❑ Since sketches cannot show how a design will work, models, mock-ups, and prototypes are made.

The Design Process: Selecting the Best Solution

- ❑ The solution to the design problem must be the best overall.
- ❑ The best solution might be a compromise between the perfect product and one people can afford.
- ❑ Managers and technical experts review the designs.
- ❑ The managers and experts consider each design from several points of view:
  - ❑ **Market.** Will the design meet needs and wants? Can it compete?
  - ❑ **Production.** Can the design be built economically? Does it fit other products the company makes?
  - ❑ **Financial.** Can the company afford the investment? Will the product make the company money?
  - ❑ **Technical.** Does the company have the engineering know-how to design and build the product?
  - ❑ **Environmental.** Will the product harm the environment? Can it be recycled?
  - ❑ **Social.** Will the product fit into the social, cultural, and religious standards of society?

Reproducible Master 14-3

The Design Process: Testing Design Solutions

- ❑ The models made earlier are tested to determine whether or not they meet the need and solve the problem.
- ❑ Working models are tested for many characteristics:
  - ❑ **Function.** Do the models meet the stated need?
  - ❑ **Performance.** Do the models work well?
  - ❑ **Endurance.** Will the models last?
  - ❑ **Operation.** Can the models be easily used?
  - ❑ **Safety.** Can the models be used without harm to people?
- ❑ The test results can lead to design changes.
- ❑ Designs that perform poorly might be abandoned.
- ❑ The designs that meet the expectations move on to the next design phase.

Reproducible Master 16-1

The Design Process: Interpreting the Solution

- ❑ Before the product can be made, the following information is needed:
  - ❑ The exact size and shape of the parts.
  - ❑ The location of all parts in the product.
  - ❑ The number and types of parts needed.
- ❑ Interpretation is done with the use of three types of communication documents:
  - ❑ Bills of materials.
  - ❑ Drawings.
  - ❑ Specification sheets.

Reproducible Master 17-1

Making an Oblique Drawing

Establish front view.

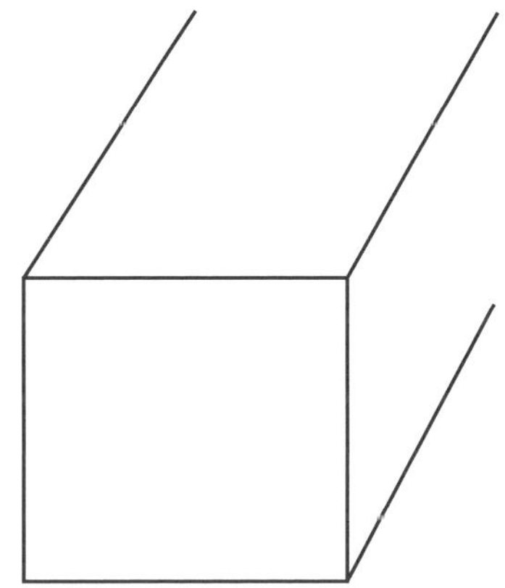

Establish side and top views.

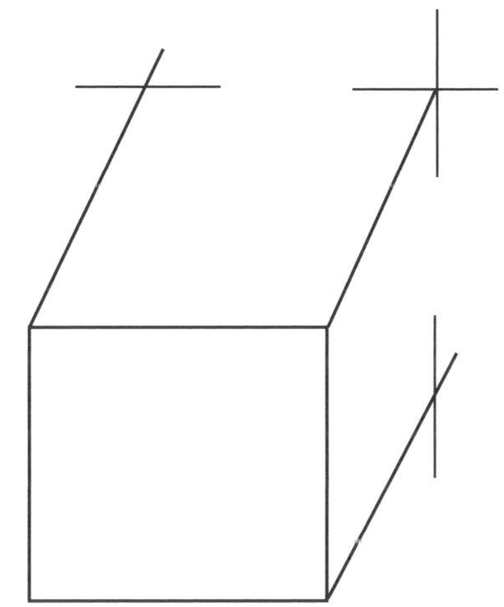

Establish depth.

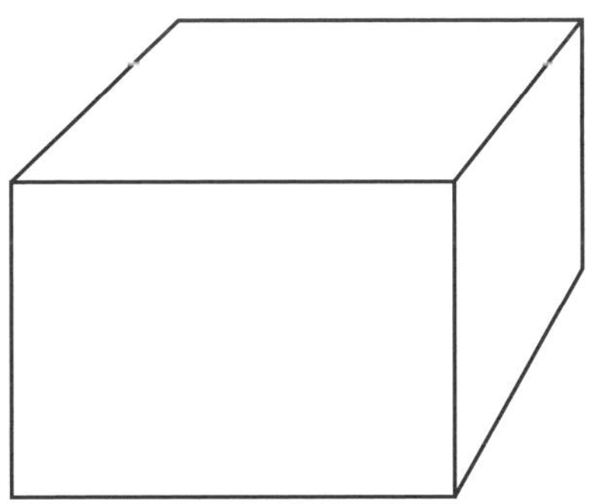

Complete block outline.

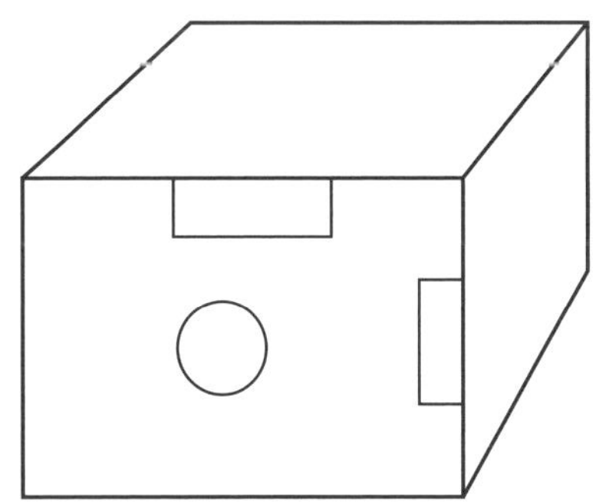

Locate features.

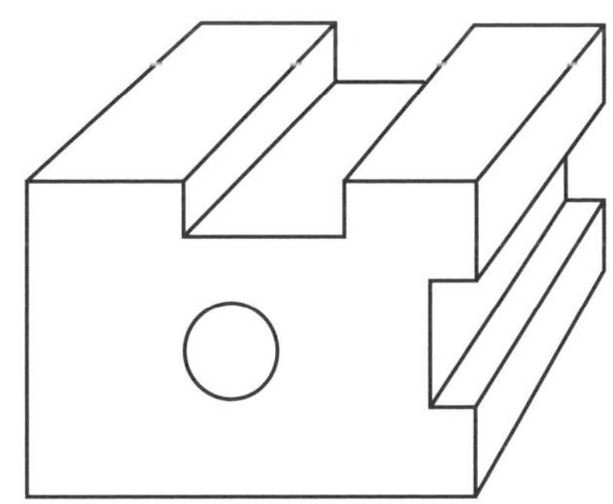

Complete drawing.

The Design Process: Presenting Solutions for Approval

- ❑ Management must approve all products before they are manufactured.
- ❑ Often, a formal presentation of the product is made.
- ❑ Written reports highlight the features of the design:
  - ❑ Outline the product and its anticipated costs.
  - ❑ List the benefits to the company.
  - ❑ Predict the expected market reaction.
  - ❑ Estimate profits.
- ❑ Usually, an oral report follows the written report, accompanied by the following media:
  - ❑ Charts.
  - ❑ Slides.
  - ❑ Transparencies.
  - ❑ Other media.

Reproducible Master 17-3

Section 4
Technological Contexts

Focus: Presentation of the agricultural and biorelated, construction, energy-conversion, information and communication, manufacturing, medical, and transportation contexts in which technology is applied.

- ❑ Chapter 19: Agricultural and Related Technology
- ❑ Chapter 20: Construction Technology
- ❑ Chapter 21: Energy-Conversion Technology
- ❑ Chapter 22: Information and Communication Technology
- ❑ Chapter 23: Manufacturing Technology
- ❑ Chapter 24: Medical Technology
- ❑ Chapter 25: Transportation Technology

Teaching Materials

Text

Chapter 19, pages 412–441
- Chapter Activities, page 438
- Test Your Knowledge, pages 438–439
- Chapter 19 Activity A—Agriculture and Biorelated Technology, page 440

Chapter 20, pages 442–473
- Chapter Activities, page 463
- Test Your Knowledge, page 464
- Chapter 20 Activity A—Construction Technology, pages 465–470

Chapter 21, pages 474–499
- Chapter Activities, page 494
- Test Your Knowledge, page 494
- Chapter 21 Activity A—Energy-Conversion Technology, pages 495–496

Chapter 22, pages 500–543
- Chapter Activities, page 536
- Test Your Knowledge, pages 536–537
- Chapter 22 Activity A—Information and Communication Technology, pages 538–540

Chapter 23, pages 544–575
- Chapter Activities, page 568
- Test Your Knowledge, pages 568–569
- Chapter 23 Activity A—Manufacturing Technology, pages 570–573

Chapter 24, pages 576–597
- Chapter Activities, page 594
- Test Your Knowledge, page 595
- Chapter 24 Activity A—Medical Technology, pages 596–597

Chapter 25, pages 598–625
- Chapter Activities, page 619
- Test Your Knowledge, page 620
- Chapter 25 Activity A—Transportation Technology, pages 621–625

Tech Lab Workbook

Chapter 19
- Chapter 19 Study Questions, pages 103–104

Chapter 20
- Student Activity 20-1—Estimating Construction Costs, pages 105–106
- Student Activity 20-2—Model-Bridge Design Contest, pages 107–108
- Chapter 20 Study Questions, page 109

Chapter 21
- Chapter 21 Activity A Worksheet—Energy-Conversion Technology, pages 111–112
- Student Activity 21-1—Energy as a Resource, pages 113–114
- Chapter 21 Study Questions, pages 115–116

Chapter 22
- Chapter 22 Activity A Worksheet—Information and Communication Technology, pages 117–121
- Student Activity 22-1—Billboard Message, page 123
- Chapter 22 Study Questions, pages 125–127

Chapter 23
- Chapter 23 Activity A Worksheet—Manufacturing Technology, pages 129–130

Teacher's Resources

Lesson Slide 25-1—Degrees of Freedom (CD only)
Lesson Slide 25-2—Transportation Processes (CD only)
Lesson Slide 25-3—Transportation Systems: Environment (CD only)
Lesson Slide 25-4—Vehicles (CD only)

Introduction

Technology is used in almost every human activity. To grow and harvest crops, process foods, communicate information, power machines and vehicles, build homes and offices, make products, transport people and goods, and treat injuries and illnesses, technology is used. These areas in which technology is used are called *contexts*. They are the areas in society in which people use technology to extend their potentials.

Objectives of the Section

At the end of Section 4, each student should be able to do the following:

- ❑ Recall the seven common technological contexts.
- ❑ Explain how people use technology to grow and harvest crops.
- ❑ Summarize the uses of the major types of constructed structures.
- ❑ Explain the major processes used to select and prepare sites, construct structures, and finish constructed projects.
- ❑ Give examples of how technology is used to convert and use energy.
- ❑ Give examples of several ways to use communication technology.
- ❑ Summarize the major steps in producing and delivering a message.
- ❑ Summarize the major types of manufacturing systems.
- ❑ Explain the major steps in developing and using a manufacturing system.
- ❑ Explain how technology is used to maintain wellness and treat illness.
- ❑ Explain how technology is used to transport people and goods.
- ❑ Give examples of the parts of a transportation system.

Suggested Schedule and Length of the Section (3–18 Weeks)

During the study of technological contexts, ask the students to complete the following assignments:

1. Read Chapters 19–25 in their textbooks and answer associated study questions.
2. Participate in presentations and discussions about the seven major technological contexts (agriculture, construction, energy-conversion, information and communication, manufacturing, medicine, and transportation).
3. Complete the scheduled laboratory activities and follow-up sheets reinforcing the study of the technological contexts.

Schedule for Section 4—9-Week and 12-Week* Classes	
Day 1	Discussion: "Technological contexts." Assign Chapter 19: "Agricultural and Related Technology" to be read outside of class. Introduce first activity (agricultural, construction, energy-conversion, or information and communication technology activity).
Day 2	Discussion: "Agricultural and biorelated technology." Work on activity.
Day 3	Work on activity. Assign Chapter 20: "Construction Technology" to be read outside of class.
Day 4	Discussion: "Construction technology." Work on activity.
Day 5	Work on activity. Assign Chapter 21: "Energy-Conversion Technology" to be read outside of class.
Day 6	Discussion: "Energy-conversion technology." Work on activity.
Day 7	Complete and summarize activity. Assign Chapter 22: "Information and Communication Technology" to be read outside of class.

Day 8	Discussion: "Information and technology." Introduce second activity (manufacturing, medical, or transportation technology activity).
Day 9	Assign Chapter 23: "Manufacturing Technology" to be read outside of class. Work on activity.
Day 10	Discussion: "Manufacturing technology." Work on activity.
Day 11	Assign Chapter 24: "Medical Technology" to be read outside of class. Work on activity.
Day 12	Discussion: "Medical technology." Work on activity.
Day 13	Assign Chapter 25: "Transportation Technology" to be read outside of class. Work on activity.
Day 14	Discussion: "Transportation technology." Work on activity.
Day 15	Complete and summarize activity. Review and summarize section.

* For a 12-week class, add another activity and spread the discussions over the time scheduled for this section.

Schedule for Section 4—18-Week and 36-Week* Classes	
Day 1	Discussion: "Technological contexts." Assign Chapter 19: "Agricultural and Related Technology" to be read outside of class. Introduce first activity (agricultural or construction technology activity).
Day 2–4	Discussion: "Agricultural and biorelated technology." Work on activity.
Day 5	Work on activity. Assign Chapter 20: "Construction Technology" to be read outside of class.
Day 6–9	Discussion: "Construction technology." Work on activity.
Day 10	Complete and summarize activity. Assign Chapter 21: "Energy-Conversion Technology" to be read outside of class.
Day 11–12	Introduce second activity (energy-conversion or information and communication technology activity). Discussion: "Energy-conversion technology."
Day 13–15	Work on activity. Assign Chapter 22: "Information and Communication Technology" to be read outside of class.
Day 16–17	Discussion: "Information and technology." Work on activity.
Day 18	Complete and summarize activity.
Day 19	Introduce third activity (manufacturing, medical, or transportation technology activity).
Day 20–21	Assign Chapter 23: "Manufacturing Technology" to be read outside of class. Work on activity.
Day 22–23	Discussion: "Manufacturing technology." Assign Chapter 24: "Medical Technology" to be read outside of class. Work on activity.
Day 24–25	Discussion: "Medical technology." Work on activity.
Day 26–27	Assign Chapter 25: "Transportation Technology" to be read outside of class. Work on activity.
Day 28–29	Discussion: "Transportation technology." Work on activity.
Day 30	Complete and summarize activity. Review and summarize section.

* For a 36-week class, use an activity for each context and spread the discussions over the time scheduled for this section.

Preparing to Teach the Section

Complete the following tasks before introducing this section to the students:

1. Study this section of the Teacher's Resource.
2. Read Chapters 19–25 in the textbook to establish the scope and sequence of the content.
3. Prepare the following 10- to 15-minute discussions and presentations:
   A. Technological contexts. (Textbook, Section 4)
   B. Agricultural and biorelated technology. (Textbook, Chapter 19)
   C. Construction technology. (Textbook, Chapter 20)
   D. Energy-conversion technology. (Textbook, Chapter 21)
   E. Information and communication technology. (Textbook, Chapter 22)
   F. Manufacturing technology. (Textbook, Chapter 23)
   G. Medical technology. (Textbook, Chapter 24)
   H. Transportation technology. (Textbook, Chapter 25)
4. Select appropriate laboratory activities from the following options:
   A. Agriculture and Biorelated Technology (Chapter 19), Textbook, page 440.
   B. Construction Technology (Chapter 20), Textbook, pages 465–470.
   C. Estimating Construction Costs (Chapter 20), Tech Lab Workbook, page 105.
   D. Model-Bridge Design Contest (Chapter 20), Tech Lab Workbook, page 107.
   E. Energy-Conversion Technology (Chapter 21), Textbook, pages 495–496.
   F. Energy as a Resource (Chapter 21), Tech Lab Workbook, page 113.
   G. Changing Potential Energy to Kinetic Energy (Chapter 21), Teacher's Resource, page 153.
   H. Information and Communication Technology (Chapter 22), Textbook, pages 538–540.
   I. Billboard Message (Chapter 22), Tech Lab Workbook, page 123.
   J. Communicating without Words (Chapter 22), Teacher's Resource, page 157.
   K. Manufacturing Technology (Chapter 23), Textbook, pages 570–573.
   L. Recipe Holder (Chapter 23), Tech Lab Workbook, page 131.
   M. Medical Technology (Chapter 24), Textbook, pages 596–597.
   N. Sports Technology (Chapter 24), Tech Lab Workbook, page 135.
   O. Transportation Technology (Chapter 25), Textbook, pages 621–625.
   P. Hovercraft Design (Chapter 25), Tech Lab Workbook, page 141.
   Q. Powerboat Design (Chapter 25), Teacher's Resource, page 158.
5. Prepare the materials and supplies for the laboratory activities.
6. Identify and construct the necessary tooling to produce the manufactured products using line-production techniques.
7. Build a suitable waterway to be used with the transportation activity (in the textbook, pages 621–625).
8. Work through the laboratory activities and student study questions presented in the textbook, Tech Lab Workbook, and Teacher's Resource.

Presenting the Section

The goal of this section is to present the societal contexts in which technology is applied. Through the laboratory work, the students will complete hands-on activities designed to present several of these contexts. Skill development is *not* a goal of these activities. To maintain a balance between content presentations and laboratory reinforcement, the following suggestions are made:

- ❑ Limit all discussions and presentations to 15 minutes. Plan each presentation carefully. Use adequate teaching materials (such as transparencies).

- ❑ Prepare materials for the laboratory activities in advance of implementing the activities. In most cases, it is not possible to have the students complete a laboratory activity for each context. The activities chosen should be broad and, whenever possible, apply more than one context. For example, a transportation activity applies energy-conversion technologies. Constructed works use manufactured components.

To maintain the schedule, the following points should be considered:

- ❑ Process the materials for the activities before introducing them.
- ❑ Build all support devices and tooling before introducing an activity.
- ❑ Produce sample products before introducing an activity.

Adjustments for Students with Special Needs and Gifted Students

For this section, 17 suggested activities have been developed. Several of them are structured activities. In these activities, the students are given a prescribed procedure to follow. Other activities are open-ended experiences developed around design briefs. Still other activities can be presented as either prescriptive or open-ended activities. All the activities lend themselves to cooperative group work. Using groups is a very effective way to adjust instruction for the differing abilities and varying interests of the students in a class. Each group can be developed to represent a mix of students. This mix includes gifted students, average students, and students with special needs. In all activities, special attention and individual help should be given to those students who require more structure. You are encouraged, however, to resist developing a high degree of structure for every activity. Generally, highly structured activities lack the level of challenge needed for many gifted students.

Evaluation Procedures

This section emphasizes using manufacturing, construction, and biorelated technology to produce products and structures. The focus is on the concepts of production, not the skills production workers use. Therefore, the students should be evaluated on the following criteria:

1. Contributions to group activity. This can be done with examination of daily logs (diaries) the students maintain, teacher observations, and student feedback.
2. Creativity and divergent thinking shown during the design and problem-solving activities.
3. Laboratory sheets, sketches, and designs completed during each activity.
4. Participation in discussions conducted during the laboratory activities.
5. Scores on quizzes and chapter questions.

Implementing the Laboratory Activities

Time has been allocated to this section for at least two activities. Classes with more time available can include more activities so the students can directly experience more technological contexts. Each activity presented is only suggested. Different activities can be substituted. The goal is to have the students apply technology in two or more societal contexts. It might be wise to consider using a structured activity for one context and a design-brief activity for another.

Procedures for Directing Each Suggested Activity

Chapter 19: Agriculture and Biorelated Technology, Textbook, page 440

This is a structured activity introducing students to food processing. They will see how an agricultural product is changed into an everyday food product. The students will make peanut butter from roasted peanuts, peanut oil, and salt. *Note:* If salted peanuts are used, omit the salt from the ingredients. At the end of the activity, the students can be treated to peanut butter and crackers.

Chapter 20: Construction Technology, Textbook, pages 465–470

Chapter 20 Activity A is a structured activity introducing students to common residential-construction practices. The activity is designed for group work. Each group follows a prescribed procedure to construct a major part of the structure. The parts are then assembled into a finished structure. At the start of the activity, divide the class into groups, as described in the activity. Show the class a completed model. Have each group carefully read the instructions for completing its part of the activity. The students should complete these steps, exercising care to follow the safety practices demonstrated during the class. The input material should be prepared before the activity is introduced. This includes scale 2 × 4s, 1 × 4s, and plywood sheets. The sizes of these materials are listed on page 465 of the textbook and are as follows:

- ❑ 8′ 2 × 4s = 1/4″ × 5/8″ × 16″.
- ❑ 8′ 2 × 6s = 1/4″ × 1″ × 16″.
- ❑ 8′ 1 × 4s = 1/8″ × 5/8″ × 16″.
- ❑ 1/4″ × 4′ × 8′ plywood = 8″ × 16″ × 6-ply poster board.
- ❑ 1/2″ × 4′ × 8′ plywood = 8″ × 16″ mat board.

Student worksheets to support this activity can be found on pages 105–106 in the Tech Lab Workbook.

Chapter 20: Estimating Construction Costs, Tech Lab Workbook, pages 105–106

"Estimating Construction Costs" helps students understand the several cost factors relating to a construction project. The students will identify and estimate the cost of land, labor, equipment, and materials for Chapter 20 Activity A. Have the students complete Student Activity 20-1 by doing the following:

1. Writing the name of the construction project: Storage shed.
2. Writing the part of the shed their group built.
3. Sketching the part of the shed they built on the grid provided.
4. Estimating the cost of the land, labor, equipment, and materials by completing the back side of the worksheet.

Chapter 20: Model-Bridge Design Contest, Tech Lab Workbook, pages 107–108

This is an open-ended activity based on a design brief. The students should be presented with the design problem and the overall procedures. Both of these elements are provided on page 107 of the Tech Lab Workbook. The bridge tester (Tech Lab Workbook, Figure 20-2B) that will be used should be shown to the class. Have the students complete the following tasks:

1. Review the various truss construction patterns.
2. Select patterns they think will be the strongest.
3. Construct models of a bridge using these patterns.
4. Test and evaluate the finished models.

Be sure the students use only the prescribed type and quantity of materials so all members of the class have an equal opportunity to design the strongest bridge.

Chapter 21: Energy-Conversion Technology, Textbook, pages 495–496, and Tech Lab Workbook, pages 111–112

Chapter 21 Activity A is a structured experiment presenting one of the many energy-conversion processes. The students should be given the appropriate supplies and equipment. The procedure for the experiment should be demonstrated. The class should then be divided into groups. Each group should complete the experiment with one color of can. The groups will share their results with the class so a comparison between color and light-to-heat energy conversion can be made.

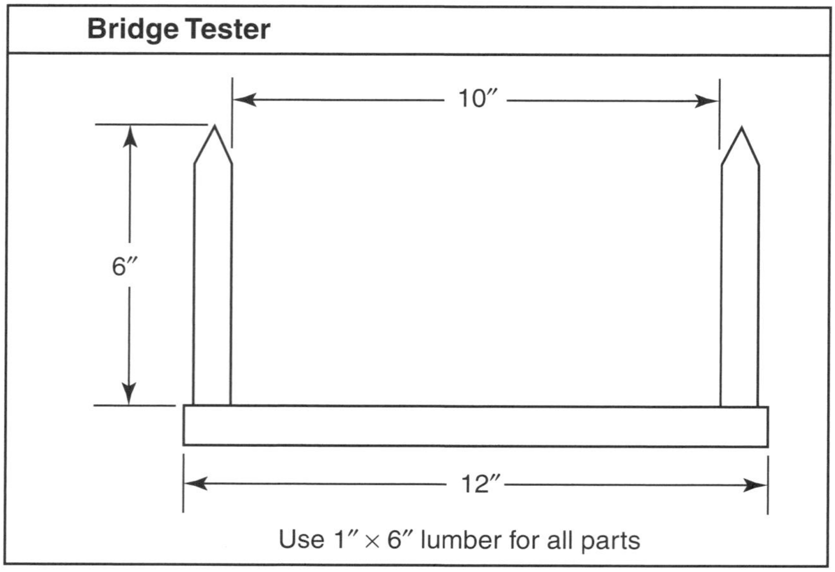

Supplies and Equipment

- ❑ A heat lamp.
- ❑ Large (42-oz.) juice cans or #10 cans (from school cafeteria).
- ❑ Spray cans of various colors of paint—black, red, blue, green, yellow, brown, and white.
- ❑ A thermometer reading up to 200°F.
- ❑ Rubber grommets with a hole the diameter of the thermometer. (Corks with a hole in the center can be substituted.)
- ❑ A data-recording sheet (included).
- ❑ A laboratory summary sheet (included).

Procedure

Preparing for the activity

Before the activity is introduced, the following steps must be completed:

1. Obtain a heat lamp for each group.
2. Construct a lamp holder (socket, switch, and bracket) for each group. This lamp holder should hold the lamp parallel to the cans and be movable to the three positions specified for the experiment.
3. Obtain juice or vegetable cans.
4. Establish the distances and times to be used in the experiment. These will vary with the heat lamp used. You will want to use times and distances showing significant results. Experiment to establish these times and distances. (Start with 12″, 18″, and 24″, for 5, 10, and 15 minutes).

Conducting the activity

1. Divide the class into groups of two to three students.
2. Assign each group a color and task, such as the following: "Determine the heat absorption of your color, in terms of the intensity of light (distance from the can to the light) and the time the light is exposed to the surface."
3. Have each group complete the following steps:
   - A. Select an appropriate can.
   - B. Locate the laboratory sheet in the Tech Lab Workbook.
   - C. Set up the experiment as demonstrated. *Note:* You should first demonstrate the proper, safe way to conduct the experiment.
   - D. Take the reading at distance #1 for each can.
   - E. Repeat steps 3C and 3D for distances #2 and #3.
   - F. Compare results of the experiment by reporting the group's results.
   - G. Record the results from other groups on the laboratory sheet.
   - H. Complete a laboratory summary sheet.

Student worksheets to support this activity can be found on pages 111–112 in the Tech Lab Workbook.

Chapter 21: Energy as a Resource, Tech Lab Workbook, pages 113–114

This activity is an open-ended, problem-solving activity challenging the students to develop an energy converter that changes wind energy into electrical energy. To carry out this activity, do the following tasks:

1. Divide the class into groups of two to three students.
2. Give each group the following basic materials:
   - ❑ A low-voltage direct current (DC) motor (available at hobby and electrical supply stores).
   - ❑ 24″ of magnet wire.
   - ❑ A low-voltage voltmeter.
3. Allow each group access to assorted materials, such as aluminum foil, dowels, thin plywood, strips of pine or balsa wood, poster board, and copper or galvanized wire.
4. Have each group complete the following steps:
   A. Sketch at least three ways the group can build the converter.
   B. Select and refine the best idea.
   C. Build a prototype of the idea.
   D. Test the prototype.
   E. Modify the prototype based on test results.
   F. Discuss ways to improve the device to make the device more efficient.

Chapter 21: Changing Potential Energy to Kinetic Energy, Teacher's Resource, page 153

This activity allows groups of students to explore how potential energy is converted into kinetic energy. To conduct this activity, you need to build a test device (pile driver) for each group of students. (See page 142.) For each device, you need the following materials:

- ❑ **Base.** One piece of 2″ × 7″ × 12″ wood.
- ❑ **Top.** One piece of 3/4″ × 2 3/4″ × 12″ wood.
- ❑ **Hammerhead.** One piece of 2 1/4″ × 4″ × 7″ wood.
- ❑ **Hammerhead face.** One piece of 1/8″ × 3/4″ × 5 3/4″ steel.
- ❑ **Guide columns.** Two 1″ D × 50″ conduits.

You also need four sizes of nails and a 2 × 4 × 6″ block for each group. The procedure for conducting the activity and a data-recording sheet are included in this Teacher's Resource.

Chapter 22: Information and Communication Technology, Textbook, pages 538–540, and Tech Lab Workbook, pages 117–121

This is a structured activity that can be used to study communication technology and custom manufacturing. The activity challenges the students to first construct simple telegraphs and then communicate messages using their communication devices. Chapter 22 Activity A contains a specific set of procedures the students are to follow. The students can develop codes to communicate messages. Students can work individually or in groups of two or three students. Half of the individuals or groups should make sending units. The other half should make receivers. To implement the activity, you need the equipment and supplies listed on page 538 of the textbook. Each student or group of students should be given a set of supplies. The students are then challenged to complete the following steps:

1. Study the plans for the sender or receiver.
2. Follow the procedures and plans to construct the device.
3. Join another student or group of students who made the other type of device to develop a complete telegraph (hardwired) communication system.
4. Develop a code to complete the message assignment on page 117 of the Tech Lab Workbook.

Student worksheets to support this activity can be found on pages 117–121 in the Tech Lab Workbook.

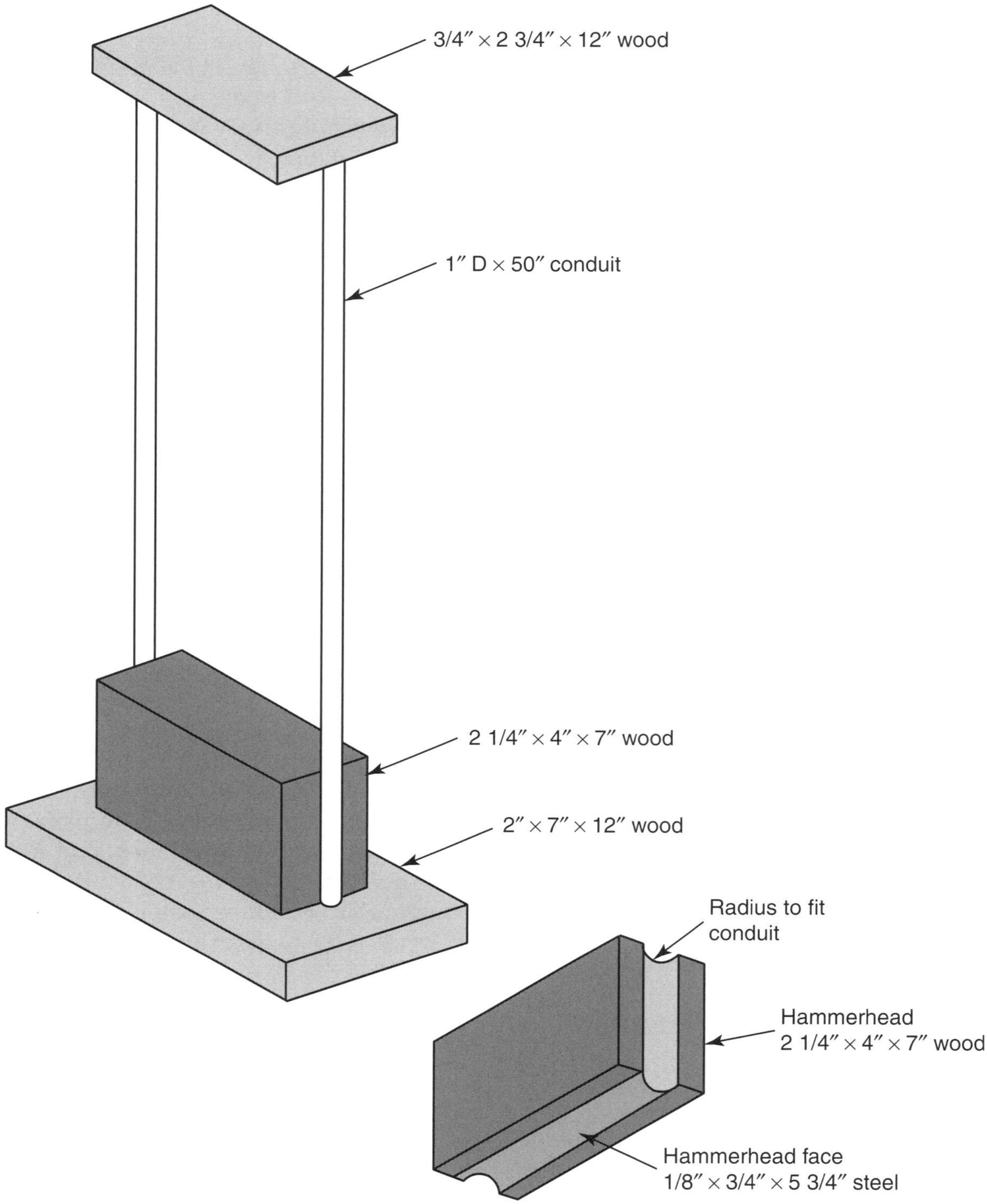

Chapter 22: Billboard Message, Tech Lab Workbook, page 123

This activity is an open-ended problem-solving activity challenging the students to develop one of two different messages:

- ❑ A billboard to promote safe bicycle riding.
- ❑ A poster to promote litter control.

To complete the activity, allow the students to work alone or in groups of two to three students. Each student or group should complete the following steps:

1. Develop a theme for the communication campaign.
2. Develop a message for the communication device supporting the campaign theme.

3. Select or develop graphics supporting the message and campaign theme.
4. Prepare a layout for the poster or billboard.
5. Produce the poster or a miniature billboard.

Chapter 22: Communicating without Words, Teacher's Resource, page 157

"Communicating without Words" is an open-ended problem-solving activity challenging the students to develop a system of communication other than the spoken word. To complete the activity, have the students work alone or in groups of two to three students. Each student or group should complete the following steps:

1. Develop a message.
2. Prepare sketches of at least three systems.
3. Select or develop graphics supporting the message.
4. Prepare a layout for the design (such as a poster or billboard).
5. Produce the system.

Chapter 23: Manufacturing Technology, Textbook, pages 570–573, and Tech Lab Workbook, pages 129–130

Chapter 23 Activity A is a structured activity challenging the students to use drawings and operation sheets to plan for and produce a manufactured product. The activity is organized around a specific set of procedures the students are to follow. The students can produce products for themselves. This, however, will be very time-consuming. A better approach is to use a teacher-designed production line to produce the product. The line should have the same three branches shown on the operation process chart on page 572 of the textbook. Each of these branches should feed an assembly line. Another approach is to divide the class into groups. Each group should work similarly to a job shop. Group 1 produces the ends. Group 2 produces the side strips. Group 3 produces the end strips. The parts from the "job shops" will be assembled into products at the end of the activity. In all cases, students should follow appropriate tool and machine procedures as demonstrated in the class. Before introducing the activity, it is necessary to prepare the input material for the line. This requires pine strips as follows: 3/4″ × 5″ × random length (end strips), 1/4″ × 3/4″ × random length (side strips), and 1/4″ × 1 1/2″ × random length (bottom strips). Remember, the goal is to introduce the students to a manufacturing system. Specific skills are not important. Rotating students through jobs is not necessary. Also, tool and material nomenclature is an inappropriate focus for this section.

This activity, similar to the rubber band–car activity in Section 2, is designed to be less structured than the game activity provided in Section 2 of the textbook. The students should be provided the drawings for the CD holder and challenged to complete the following steps:

1. Develop a bill of materials for the product.
2. Prepare a procedure sheet for building the product.
3. List any restrictions they face, such as CD-case size, number of CDs the holder will hold, and cost of the solution.
4. Build a prototype of the holder.
5. Discuss the procedures they used, ways to build the product more efficiently, and other similar topics.

Chapter 23: Recipe Holder, Tech Lab Workbook, pages 131–132

This activity is an open-ended problem-solving activity challenging the students to develop a manufacturing system for a simple product—a recipe holder. To complete the activity, divide the class into groups of three or four students to plan the system. Each group should complete the following steps:

1. Prepare a bill of materials for the product.
2. Prepare a procedure sheet for each of the two manufactured parts—the base and upright.
3. Highlight steps requiring tooling to maintain tolerance.
4. Check places where the students think inspection (quality control) is needed.

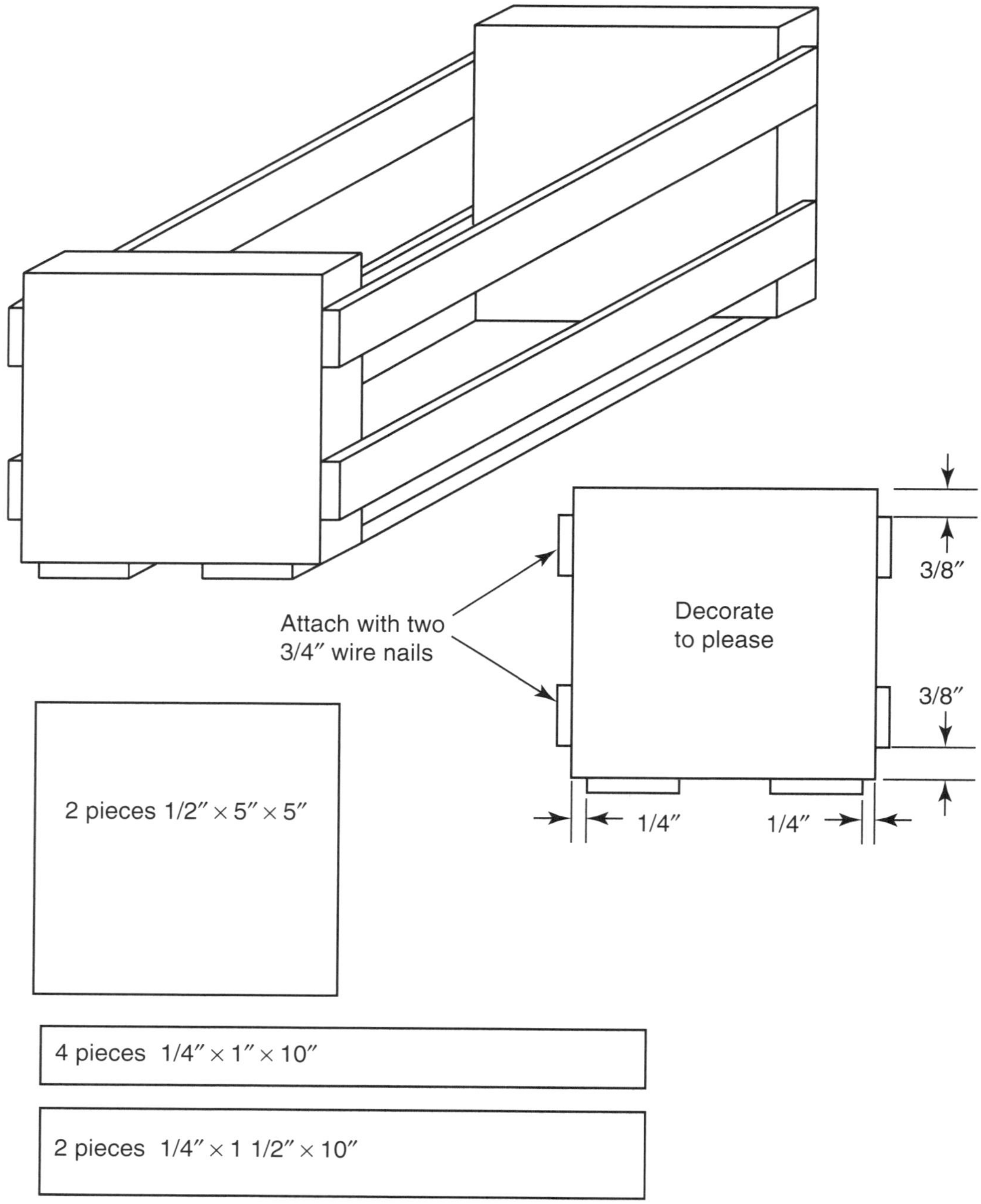

Select the best manufacturing system by choosing one group's work or integrating several groups' ideas. Build the necessary tooling and inspection gages. Manufacture the product.

Chapter 24: Medical Technology, Textbook, pages 596–597

Chapter 24 Activity A is an open-ended activity based on a design brief. The students should be presented with the design problem and the overall procedures. Both of these elements are provided on pages 596–597 of the textbook. The students are challenged to design a device allowing a person to pick objects up without bending over. This device will help people with arthritis, back injuries, and other conditions that make bending over difficult. The students should use the design process presented in Section 3 to complete this activity. They can work individually or in small groups to complete the activity.

Chapter 24: Sports Technology, Tech Lab Workbook, page 135

This activity is a design brief in which the students will use their creativity to address a specific design challenge. To conduct this activity, have the students do the following:

1. Select sports they play or with which they are familiar.
2. Identify pieces of sports equipment that would make the sports safer to play.
3. Sketch designs for the equipment on the grid provided on the activity worksheet.
4. Present their designs to the class.

Chapter 25: Transportation Technology, Textbook, pages 621–625, and Tech Lab Workbook, pages 139–140

This is a structured activity introducing students to vehicle design. They will develop and test boat hulls for effectiveness. To implement this activity, divide the class into groups of two to four students. Provide each group with the supplies listed with the activity on pages 621–622 of the textbook. Have each group carefully follow the procedure listed in their textbooks to build boat hulls, sails, and keels. The groups should test their hulls in a waterway.

Chapter 25: Hovercraft Design, Tech Lab Workbook, pages 141–142

This is a structured activity allowing students to construct and test an air-cushion vehicle (ACV). The students should be divided into groups of two to three students. Each group should be provided the supplies listed with the activity. The groups should then be challenged to complete the following steps:

1. Read the procedures for constructing the ACV.
2. Construct their vehicles.
3. Test the vehicles.
4. Suggest ways to modify the vehicles to improve their performances.

Chapter 25: Powerboat Design, Teacher's Resource, page 158

"Powerboat Design" is based on a design brief. The students should be presented with the design problem. To meet the design challenge, they should be provided with the following supplies:

- ❑ One piece of 1 1/2" × 5" × 9" foam block.
- ❑ One rubber band.
- ❑ Assorted dowels, sheet metal, balsa wood, nails, and pins.

The students must solve the problem by completing the following steps:

1. Prepare sketches of at least three designs.
2. Select, refine, and dimension the most promising design.
3. Build a prototype of the design.
4. Test the design.
5. Modify the design to make it more efficient.

Student worksheets to support this activity can be found on pages 111–119 in this Teacher's Resource.

Answer Key for Section 4

Chapter 19—Agricultural and Related Technology

Test Your Knowledge, Textbook, pages 438–439

1. Agriculture is growing and harvesting crops and raising livestock. This growing, harvesting, and raising is important because it produces the food and natural fibers needed to maintain life.
2. Crop production and animal husbandry.
3. grains
4. forage crops
5. planting, growing, harvesting, storing
6. Any three of the following: cattle ranches, dairies, swine farms, horse farms, and poultry farms. Explanations will vary.
7. aquaculture
8. biotechnology
9. gene splicing
10. A. Refrigeration and freezing.
    B. Dehydration.

C. Chemical preservation.
D. Irradiation.

11. To reproduce some facets of the natural environment so scientists can study agricultural processes and systems as they would be useful to biological ecosystems.
12. To help ensure that food is safe to eat.

Study Questions, Tech Lab Workbook, pages 103–104

1. technology
2. True
3. Crop production, animal husbandry (livestock production).
4. Grains, vegetables, fruits, nuts, forage crops, and nonfood.
5. E. Power or pulling equipment.
6. G. Tilling equipment.
7. A. Planting equipment.
8. C. Pest-control equipment.
9. B. Irrigation equipment.
10. A. Harvesting equipment.
11. F. Storage equipment.
12. Cattle ranches, dairies, swine farms, horse farms, poultry farms.
13. hydroponics
14. fish, shellfish, and aquatic plants
15. Biotechnology
16. Refrigeration and freezing, canning, dehydrating, chemical preservation, irradiation.
17. artificial ecosystem
18. quality control

Chapter 20—Construction Technology

Test Your Knowledge, Textbook, page 464

1. constructed work
2. Can be used as a grocery or clothing store.—B. Commercial.
   Contains spaces for living, food preparation, and sleeping.—A. Residential.
   Can be called a *church*, *synagogue*, or *mosque*.—E. Religious.
   Can be used as a jail or firehouse.—D. Public.
   Can be apartments.—A. Residential.
   Can be used as factories.—C. Industrial.
   Used by businesses providing services or selling goods.—B. Commercial.
3. A constructed work supporting the public interest or other technological activities.
4. need
5. foundation
6. Floor, walls, and roof.
7. To keep out the weather, protect people and goods, and give privacy.
8. mechanical systems
9. insulation
10. landscaping

Study Questions, Tech Lab Workbook, page 109

1. Residential, commercial, industrial, public, and religious.
2. taxes
3. civil
4. drawings
5. boundaries
6. foundation
7. Floors, walls, and a roof.
8. studs
9. plumbing
10. Insulation
11. Landscaping

Chapter 21—Energy-Conversion Technology

Test Your Knowledge, Textbook, page 494

1. work
2. alternator (electric generator)
3. solid fuel, liquid fuel
4. heat
5. Batteries convert chemical energy into electrical energy by producing a current with the use of an anode and a cathode.
6. Fuel cells convert chemical energy into electrical energy by producing a current from fuel and oxygen. The fuel loses electrons to a carbon plate. This plate becomes negatively charged. The oxygen collects electrons from a second plate, leaving the plate with a positive charge.
7. photovoltaic cell
8. passive solar systems, active solar systems
9. passive

10. Converts liquids into another form of energy.—E. Fluid converter.
    Converts the Sun's energy into another form.—D. Solar converter.
    Converts heat energy into another form.—B. Thermal converter.
    Converts kinetic energy into another form.—A. Mechanical converter.
    Converts energy in the molecular structure of a material into another form.—C. Chemical converter.
    Examples will vary. Some acceptable answers are a wind turbine (fluid), a silicon semiconductor (solar), a heat engine (thermal), an electric generator (mechanical), and a battery (chemical).
11. Wind turbines
12. Water turbines
13. mechanical, fluid, electrical
14. step-up transformers
15. Conserving energy is important because the rate at which energy is being used in the world is increasing and most of the fuels we use are limited in supply.

Study Questions, Tech Lab Workbook, pages 115–116

1. energy
2. Mechanical, thermal, chemical, solar, and fluid converters.
3. mechanical
4. alternating
5. thermal
6. jet
7. Solid fuel and liquid fuel.
8. Batteries
9. fuel cell
10. Solar cells and solar heating systems.
11. boat
12. mechanical
13. energy conservation

Chapter 22—Information and Communication Technology

Test Your Knowledge, Textbook, pages 536–537

1. Data that has been sorted and arranged.
2. The act of exchanging ideas, information, and opinions.
3. Answers will vary. Some acceptable answers are speaking, writing, reading, drawing, and taking photographs.
4. The technical equipment used.
5. C. Listening to music on a CD player.
   F. Videotaping a school play.
   G. Drawing a design for a poster.
6. Telecommunication systems, audio and video recording systems, computer (Internet) systems, printing systems, photographic systems, and drafting systems.
7. Listening to a newscast on the radio.—A. People-to-people communication.
   Clock chimes sounding the hour.—C. Machine-to-people communication.
   Setting the alarm on a clock radio.—B. People-to-machine communication.
   A thermostat controlling the operation of a furnace.—D. Machine-to-machine communication.
8. Encoding is changing the form of the message into one the receiver will understand. Transmitting is sending the message. Receiving is arriving at the desired location. Decoding is putting meaning to the message so the message can be understood.
9. Designing the message, preparing to produce the message, producing the message, and delivering the message.
10. hardware, software

Study Questions, Tech Lab Workbook, pages 125–127

1. B. Information.
2. B. Information.
3. A. Data.
4. communication
5. Graphic communication and wave (electronic) communication.
6. communication technology
7. graphic communication
8. electronic
9. carrier
10. telecommunication
11. hypertext-markup language
12. D. Relief.
13. C. Offset lithography.

14. B. Intaglio.
15. E. Screen printing.
16. A. Electrostatic.
17. People-to-people, people-to-machine, machine-to-people, machine-to-machine.
18. A. Transmitter.
    B. Carrier.
    C. Receiver.
19. encoding
20. transmitting
21. decoding
22. A. H
    B. S
    C. S
    D. H
    E. H
    F. S
23. B. Central processing unit.
24. A. Input device.
25. D. Output device.
26. C. Memory.
27. A. Input device.
28. C. Memory.
29. D. Output device.
30. B. Central processing unit.

Chapter 23—Manufacturing Technology

Test Your Knowledge, Textbook, pages 568–569

1. raw materials, products
2. A. primary processing
   B. secondary processing
3. Cuts materials.—B. *Mechanical processing*.
   Heats materials to change them.—
   C. *Thermal processing*.
   Uses chemicals to make new materials.—
   A. *Chemical processing*.
   Examples will vary.
4. Casting and molding, forming, separating, conditioning, assembling, and finishing.
5. A. Casting and molding.
   B. Forming.
   C. Separating.
   D. Conditioning.
   E. Assembling.
   F. Finishing.
6. A. One person makes an entire product. In another instance, a few very special products are designed for the customer.
   B. Products are made in small batches. They travel from operation to operation in a batch.
   C. Parts move down a manufacturing line to an assembly line. They are put together to form the finished products on the assembly line.
   D. Small lots are produced. Continuous manufacturing actions, however, are used.
7. True
8. A. A device that holds a part while it is being manufactured.
   B. Designed to make manufacturing more efficient.
   C. Sometimes a special shape for forming the material.
9. Setting goals.—Planning.
   Dividing tasks into jobs.—Organizing.
   Assigning jobs and supervising workers.—Actuating.
   Comparing the results to the plan.—Controlling.
10. People, materials, and machines (equipment).
11. advertising

Study Questions, Tech Lab Workbook, pages 133–134

1. raw
2. primary
3. secondary processing
4. Mechanical, thermal, and chemical processes.
5. F. Finishing.
6. A. Casting and molding.
7. E. Assembling.
8. B. Forming.
9. C. Separating.
10. B. Forming.
11. D. Conditioning.
12. F. Finishing.
13. C. Separating.
14. A. Casting and molding.
15. Custom, intermittent and batch, continuous, and flexible manufacturing.

16. A. Selecting operations.
    B. Sequencing operations.
    C. Selecting equipment.
    D. Arranging equipment.
    E. Designing tooling.
    F. Controlling quality.
    G. Testing the system.
17. Flow process
18. tooling
19. inspection
20. D. Controlling.
21. C. Actuating.
22. A. Planning.
23. B. Organizing.
24. C. Actuating.
25. Advertising

Chapter 24—Medical Technology

Test Your Knowledge, Textbook, page 595

1. Tiny blood vessels.
2. Answers will vary.
3. prevent, diagnose, treat
4. It is better to prevent an injury or a disease than to try to cure one after it occurs.
5. Answers will vary. Some acceptable answers are wellness and vaccination.
6. Answers will vary. Some acceptable answers are treadmills, stationary bikes, stair climbers, rowing machines, and home gyms.
7. equipment, personal protection, venues
8. People are systematically vaccinated through a series of shots to prevent disease.
9. vaccines
10. diagnosis
11. treatments
12. X rays
13. magnetic
14. surgery
15. Drugs

Study Questions, Tech Lab Workbook, pages 137–138

1. To prevent, diagnose, and treat illnesses and injuries.
2. Wellness
3. Nutrition and diet, environment, stress management, and physical fitness.
4. B. Aerobic exercise.
5. A. Anaerobic exercise.
6. A. Anaerobic exercise.
7. B. Aerobic exercise.
8. B. Aerobic exercise.
9. Prevent illnesses, prevent injuries.
10. A vaccine
11. Diagnosis and treatment.
12. D. X-ray machine.
13. B. Magnetic resonance imaging (MRI).
14. C. Ultrasound.
15. invasive diagnostic technology
16. Therapeutic radiology
17. surgery
18. Drugs
19. Vaccines

Chapter 25—Transportation Technology

Test Your Knowledge, Textbook, page 620

1. The wheel was invented about 5000 years ago. This device is important because it laid the foundation for new land-transportation systems.
2. C. Moving people and goods from one place to another.
3. Flying on American Airlines to Chicago.—C. Commercial transportation.
   Riding a bicycle to school.—A. Personal transportation.
   Riding in the family car on a vacation.—A. Personal transportation.
   Riding in a taxi to the train station.—C. Commercial transportation.
   Riding the city bus to work.—B. Public transportation.
4. scheduling
5. storing
6. land, water, air, space
7. fixed path, variable path
8. Answers will vary.
9. The vehicle carries the people or cargo. The support facilities provide the pathway and terminals, as well as life, legal, operational, maintenance, and economic support for safe and efficient transportation.
10. structure, propulsion, suspension, guidance, control

11. support
12. Answers will vary.

Study Questions, Tech Lab Workbook, pages 143–144

1. transportation
2. A. Three.
   B. Two.
   C. Two.
   D. One.
   E. One.
   F. Three.
3. public
4. personal
5. Routing, scheduling, loading, moving, unloading, and storing.
6. fixed-path systems, variable-path systems
7. Unit
8. shipping lanes
9. intermodal transportation
10. support
11. E. Control.
12. C. Suspension.
13. A. Structure.
14. E. Control.
15. B. Propulsion.
16. D. Guidance.
17. hovercraft
18. support systems

Captions for Reproducible Masters

- ❑ **Reproducible Master 19-1.** Biotechnology's Earmarks—Some of the main characteristics of biotechnology are listed.
- ❑ **Reproducible Master 20-1.** Types of Civil Structures—Civil structures can benefit transportation, communication, and energy conversion.
- ❑ **Reproducible Master 20-2.** Steps in Constructing a Structure—The eight major steps in constructing any structure are listed.
- ❑ **Reproducible Master 20-3.** The Planning and Management of Construction Projects—There are many steps involved in planning and managing a construction project.
- ❑ **Reproducible Master 20-4.** The Many Tasks in Building a Structure—Building a structure requires the completion of several construction tasks. These tasks include installing systems, finishing the structure, and servicing the structure.
- ❑ **Reproducible Master 21-1.** Energy Conversion—Many types of energy are converted into other forms at a steam power plant.
- ❑ **Reproducible Master 22-1.** Types of Communication Systems—There are four main types of communication systems.
- ❑ **Reproducible Master 22-2.** The Communication Process—The communication process consists of six major steps.
- ❑ **Reproducible Master 22-3.** Creating Media Messages—There are four main parts in the process of creating media messages.
- ❑ **Reproducible Master 22-4.** Steps Used to Produce a Message for Print Communication—The major steps in each of the four parts in the process of creating media messages are listed.
- ❑ **Reproducible Master 22-5.** Designing a Message—The major steps in the design of printed graphic media messages and telecommunication media messages are listed.
- ❑ **Reproducible Master 22-6.** Preparing to Produce a Message—The major steps in the preparation for producing printed graphic media messages and telecommunication media messages are listed.
- ❑ **Reproducible Master 22-7.** Producing a Message—The major steps in the production of printed graphic media messages and telecommunication media messages are listed.
- ❑ **Reproducible Master 22-8.** Delivering a Message—The major steps in the delivery of printed graphic media messages and telecommunication media messages are listed.
- ❑ **Reproducible Master 22-9.** Parts of a Computer—The main parts of a computer are labeled.

- ❑ **Reproducible Master 22-10.** Communication Guidelines—Guidelines for different types of communication, including written, video, PowerPoint, Web-based, and desktop publishing are given.
- ❑ **Reproducible Master 23-1.** Secondary Manufacturing Processes—The six secondary manufacturing processes are shown.
- ❑ **Reproducible Master 24-1.** Technology and Medicine—The three main functions of technology in medicine are listed.
- ❑ **Reproducible Master 24-2.** Wellness—Wellness is a state of personal well-being.
- ❑ **Reproducible Master 25-1.** Vehicle Systems: Structure—The parts of a vehicle's structure are shown.
- ❑ **Reproducible Master 25-2.** Vehicle Systems: Propulsion—The functions of a vehicle's propulsion system are listed.
- ❑ **Reproducible Master 25-3.** Vehicle Systems: Suspension—The components of a vehicle's suspension system are listed.
- ❑ **Reproducible Master 25-4.** Vehicle Systems: Guidance—The types of information a vehicle's guidance system provides are listed.
- ❑ **Reproducible Master 25-5.** Vehicles and Their Systems—A chart for students to fill out when learning about vehicle systems.

Captions for Lesson Slides (CD Only)

- ❑ **Lesson Slide 21-1.** Energy Converters—One form of energy can be changed into a different form of energy.
- ❑ **Lesson Slide 22-1.** Graphic Communication—Graphic communication is information communicated using drawings, pictures, graphs, photographs, or words on flat surfaces.
- ❑ **Lesson Slide 22-2.** Types of Communication—The four distinct types of communication for which communication technology is used.
- ❑ **Lesson Slide 22-3.** Communication Process—The basic process has four major steps.
- ❑ **Lesson Slide 22-4.** Parts of a Computer—Shown are the four basic parts of a computer system.
- ❑ **Lesson Slide 23-1.** Producing Finished Products—There are six major secondary-processing activities.
- ❑ **Lesson Slide 23-2.** Casting and Molding—Casting is a process in which material is made into a liquid, poured into a mold, and hardened in the shape of the mold's cavity.
- ❑ **Lesson Slide 23-3.** Forming—Forming is a process in which material is squeezed or stretched into the desired shape.
- ❑ **Lesson Slide 23-4.** Separating—Separating is a process in which material is shaped by removing excess stock.
- ❑ **Lesson Slide 23-5.** Conditioning—Conditioning is a process in which a material's internal structure is altered and improved.
- ❑ **Lesson Slide 23-6.** Assembling—Assembling is a process in which parts are put together to make a product.
- ❑ **Lesson Slide 23-7.** Finishing—Finishing is a process in which a material's surface is protected and beautified.
- ❑ **Lesson Slide 25-1.** Degrees of Freedom—Vehicles have degrees of freedom to move people and cargo.
- ❑ **Lesson Slide 25-2.** Transportation Processes—Transportation involves a number of different processes that are vital for the whole transportation system to operate efficiently. They can be used alone or in a variety of combinations to transfer cargo and people.
- ❑ **Lesson Slide 25-3.** Transportation Systems: Environment—Transportation systems operate in many different environments.
- ❑ **Lesson Slide 25-4.** Vehicles—Transportation vehicles are made up of five subsystems that must operate together for a system to work efficiently.

Name ______________________ Date __________

Period ______________________ Score __________

Chapter 21 Alternative Activity

Changing Potential Energy to Kinetic Energy

Introduction

Energy is a critical part of life. This part of life is defined as the capacity to do work. Two basic types of energy are potential energy (energy of position) and kinetic energy (energy of motion). An object is said to have potential energy when it is at rest (for example, a book on a table). If the book falls, potential energy turns into kinetic energy. The book (matter) then has the ability to do work. Water falling onto a waterwheel is one way people have harnessed kinetic energy. Today, water is still a major source of energy. Water behind a hydroelectric dam is potential energy. As it falls through long pipes to the turbines at the powerhouse, it becomes kinetic energy. Wood, coal, and oil have potential energy that becomes kinetic energy during combustion. In this activity, you will use a small pile driver to experiment with the concept of converting energy from potential to kinetic.

Equipment and Materials

- ❑ A pile driver.
- ❑ One piece of 2 × 4 (any grade).
- ❑ A 1 1/2″ wire nail.
- ❑ Assorted other nails (preferably heavier than the 1 1/2″ wire nail, but no longer).
- ❑ A small hammer.
- ❑ A ruler.
- ❑ Masking tape.
- ❑ Eye protection.

Procedure

1. Obtain a pile driver.
2. Place the 2 × 4 on the base of the pile driver.
3. Use masking tape to mark 1/3 and 2/3 of the way up the column. When the middle of the pile driver hammer is at the 1/3 mark, it should be 1/3 the distance from the 2 × 4 to the top of the column.
4. Tap a 1 1/2″ wire nail in the 2 × 4 so the nail stands up. Do not hammer the nail into the wood more than necessary.

Alternate Activity *(continued)* Name ____________________

5. Measure the height of the head of the nail from the 2 × 4. Record this starting height on the data sheet.

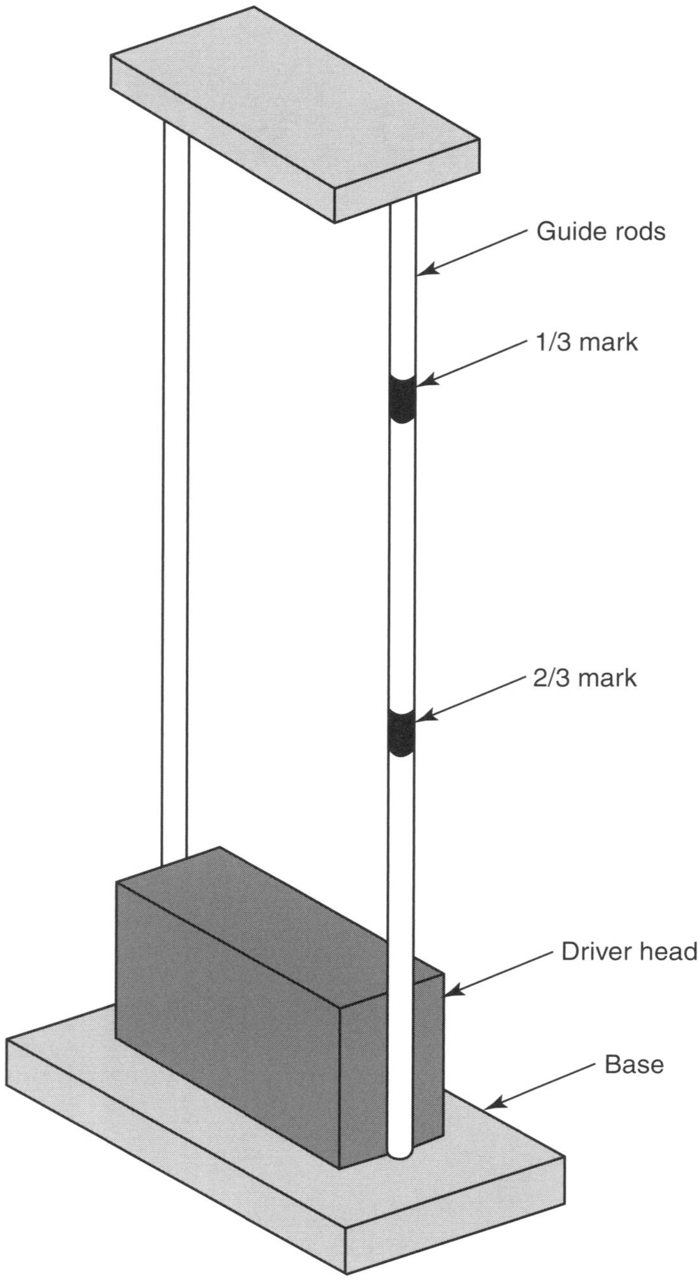

Pile Driver Test Unit

6. Raise the middle of the pile driver hammer to the 1/3 mark on the column (potential energy). Drop the hammer so it strikes the nail (kinetic energy). Record the time it takes for the hammer to strike the nail.
7. Remove the 2 × 4 from the pile driver. Measure the distance from the head of the nail to the wood. Record the information on the data sheet. Remove the nail.
8. Place a new nail in the 2 × 4. (The same nail can be used, if it is not bent.) The height of the head should be the same as in step 5.

Alternate Activity *(continued)* Name ________________________

9. Raise the pile driver hammer to the 2/3 mark (potential energy). Drop the hammerhead (kinetic energy) so it strikes the nail. Record the time it takes for the hammer to strike the nail.
10. Remove the 2 × 4 from the pile driver. Measure the distance from the head of the nail to the wood. Record the information on the data sheet. Remove the nail.
11. Place a new nail in the 2 × 4. (The same nail can be used, if it is not bent.) The height of the head should be the same as in step 5.
12. Raise the pile driver hammer to its full height (potential energy). Drop the hammer (kinetic energy) so it strikes the nail. Record the time it takes for the hammer to strike the nail.
13. Remove the 2 × 4 from the pile driver. Measure the distance from the head of the nail to the wood. Record the information on the data sheet. Remove the nail.
14. Repeat the procedure using different-size nails. Be sure to record all results for comparison.

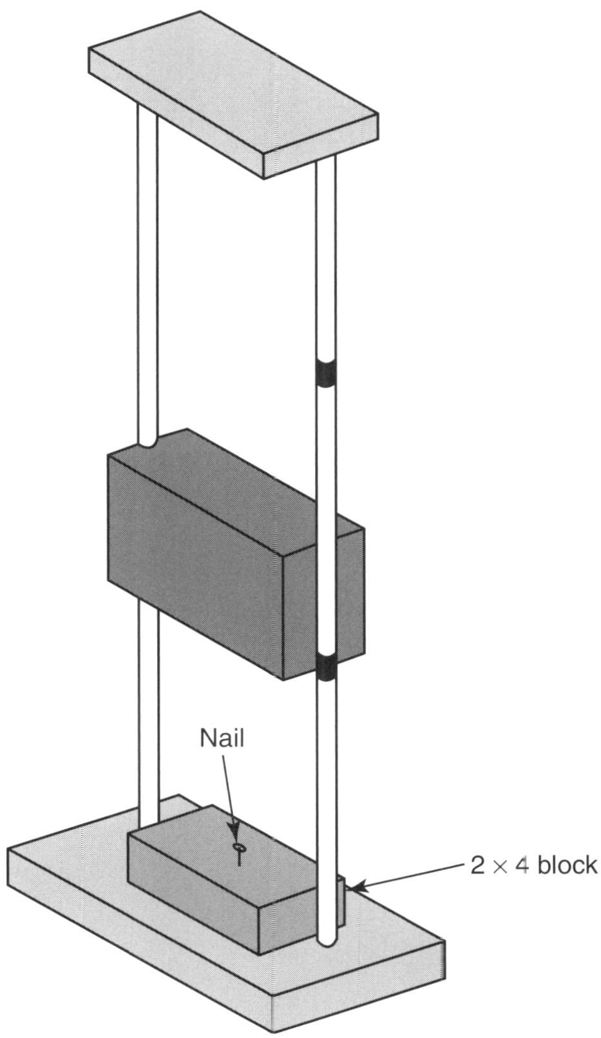

Experiment Setup

Alternate Activity *(continued)* Name ____________________

Energy Data Sheet

On the chart below, record the data from each test for each size of nail.

Nail Size	Starting Height of Nail	1/3 Drop	Time	2/3 Drop	Time	Full Drop	Time

1. Calculate the potential energy and kinetic energy for the full drop of the first two nails using the following formulas:

$$v = \frac{s}{t} \qquad \left(\text{velocity} = \frac{\text{distance}}{\text{time}}\right)$$

$$PE = mh \qquad (\text{potential energy} = \text{mass} \times \text{height})$$

$$KE = \frac{1}{2}mv^2 \; (\text{kinetic energy} = 1/2 \text{ mass} \times \text{velocity squared})$$

Nail #1 Potential energy ______ Kinetic energy ______

Nail #2 Potential energy ______ Kinetic energy ______

2. What conclusions can you make about force, relative to distance?
3. List three objects showing the characteristics of potential energy.

   A. ______

   B. ______

   C. ______
4. List three objects showing the characteristics of kinetic energy.

   A. ______

   B. ______

   C. ______

Name ____________________ Date __________

Period ____________________ Score __________

Chapter 22 Alternative Activity

Communicating without Words

The Challenge

Develop a system allowing you to communicate without words.

Background

People need to communicate. They are often, however, far apart. Many times, they cannot hear one another. For this reason, communication technology was developed. Many of these systems do not rely on "hearing" the message. They use signs and symbols to communicate. Below are some examples of ways people communicate without hearing words. The system you develop must be able to deal with all words in the English language and use technology. This communication cannot simply be body signals. There must be technology (devices people design and make) involved.

Smoke Signals

Symbols

Morse Code

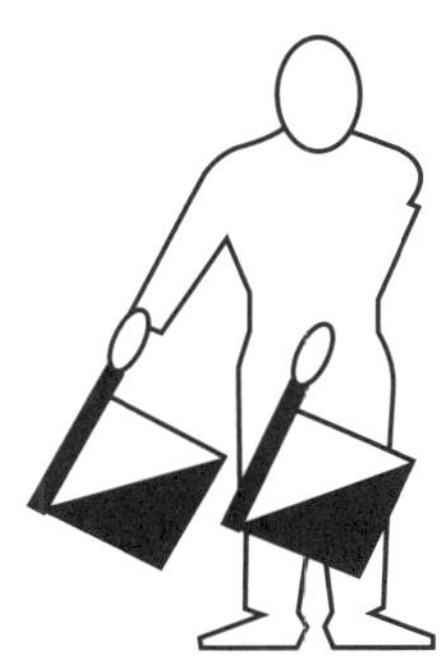

Semaphore

Design Brief

Develop your communication system using any or all of the following materials:

- ❑ 1/4" dowels.
- ❑ Colored paper.
- ❑ Colored pencils.
- ❑ Felt-tip markers.
- ❑ Paper clips.
- ❑ Poster board.
- ❑ String.

Name ______________________ Date __________

Period _____________________ Score __________

Chapter 25 Alternative Activity

Powerboat Design

Design Brief

Design a rubber band-powered boat that will carry a specified weight along a canal. The boat carrying the load the farthest with the standard rubber band as the only means of propulsion will be considered the best design.

Specifications

The maximum measurements are as follows:

- ❑ Length _____
- ❑ Width _____
- ❑ Height _____
- ❑ Specified load _____ oz.

Design

Biotechnology's Earmarks

Biotechnology: the industrial application of biological systems to process materials.

- ❑ Biotechnology uses a natural biological agent to transform materials.
- ❑ Generally, the agent is a microorganism.
- ❑ Agents are usually enzymes, animal cells, or plant cells.
- ❑ Biotechnological processes produce both goods and services.
- ❑ Some goods include the following:
  - ❑ Beverages.
  - ❑ Processed foods.
  - ❑ Methane.
  - ❑ Extracted metals.
- ❑ Services are generally treatment and purification of wastes.

Reproducible Master 19-1

Types of Civil Structures

- ❑ Transportation structures.
  - ❑ Roads.
  - ❑ Runways.
  - ❑ Rail lines.
  - ❑ Bridges.
  - ❑ Pipelines.
- ❑ Communication structures.
  - ❑ Broadcast towers.
  - ❑ Satellite dishes.
- ❑ Energy structures.
  - ❑ Dams.
  - ❑ Power lines.
  - ❑ Substations.

Steps in Constructing a Structure

1. Preparing to build.
2. Preparing the site.
3. Setting the foundation.
4. Building the framework.
5. Enclosing the structure.
6. Installing utilities.
7. Finishing the interior.
8. Completing the site.

Reproducible Master 20-2

The Planning and Management of Construction Projects

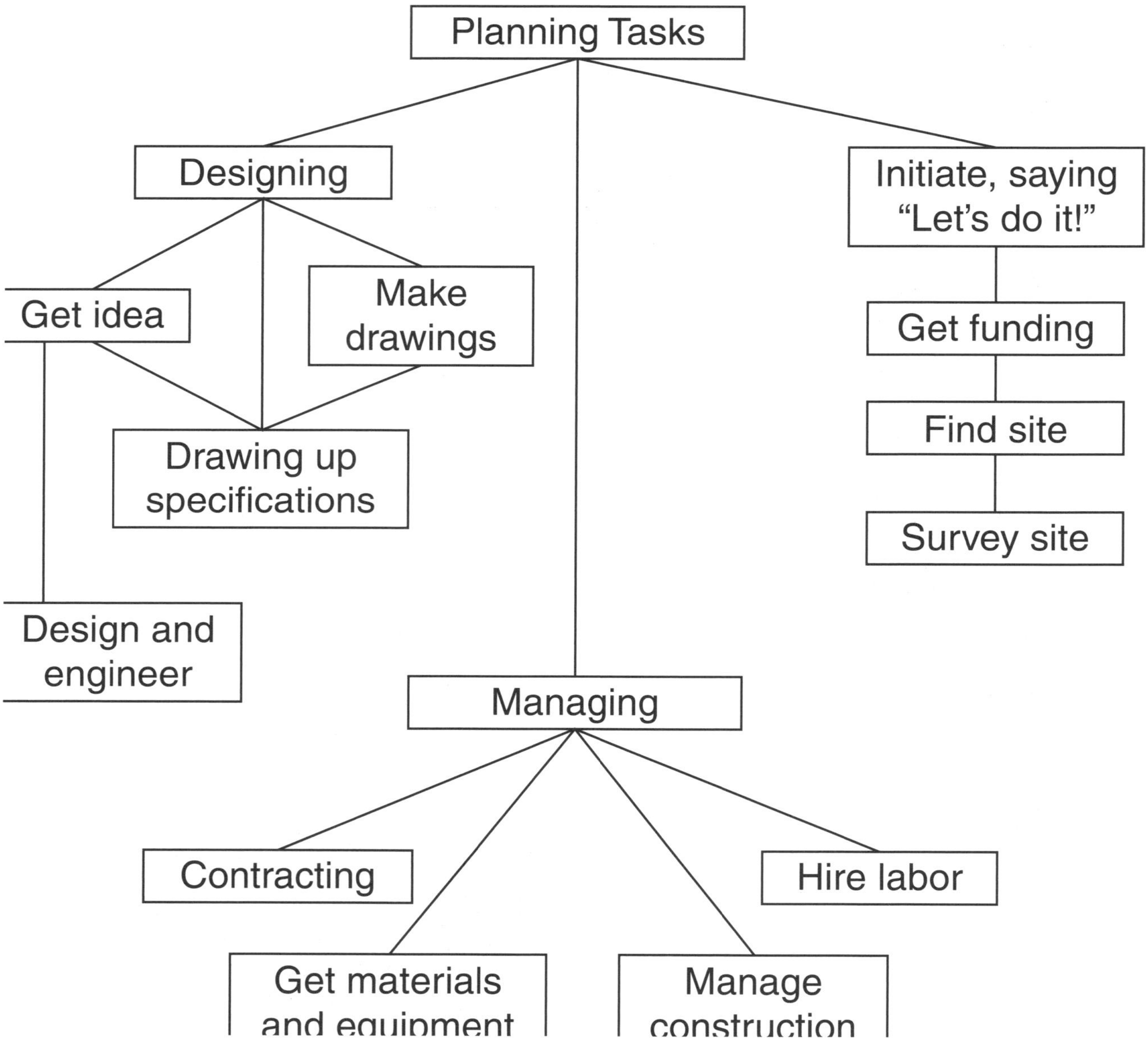

Reproducible Master 20-3

The Many Tasks in Building a Structure

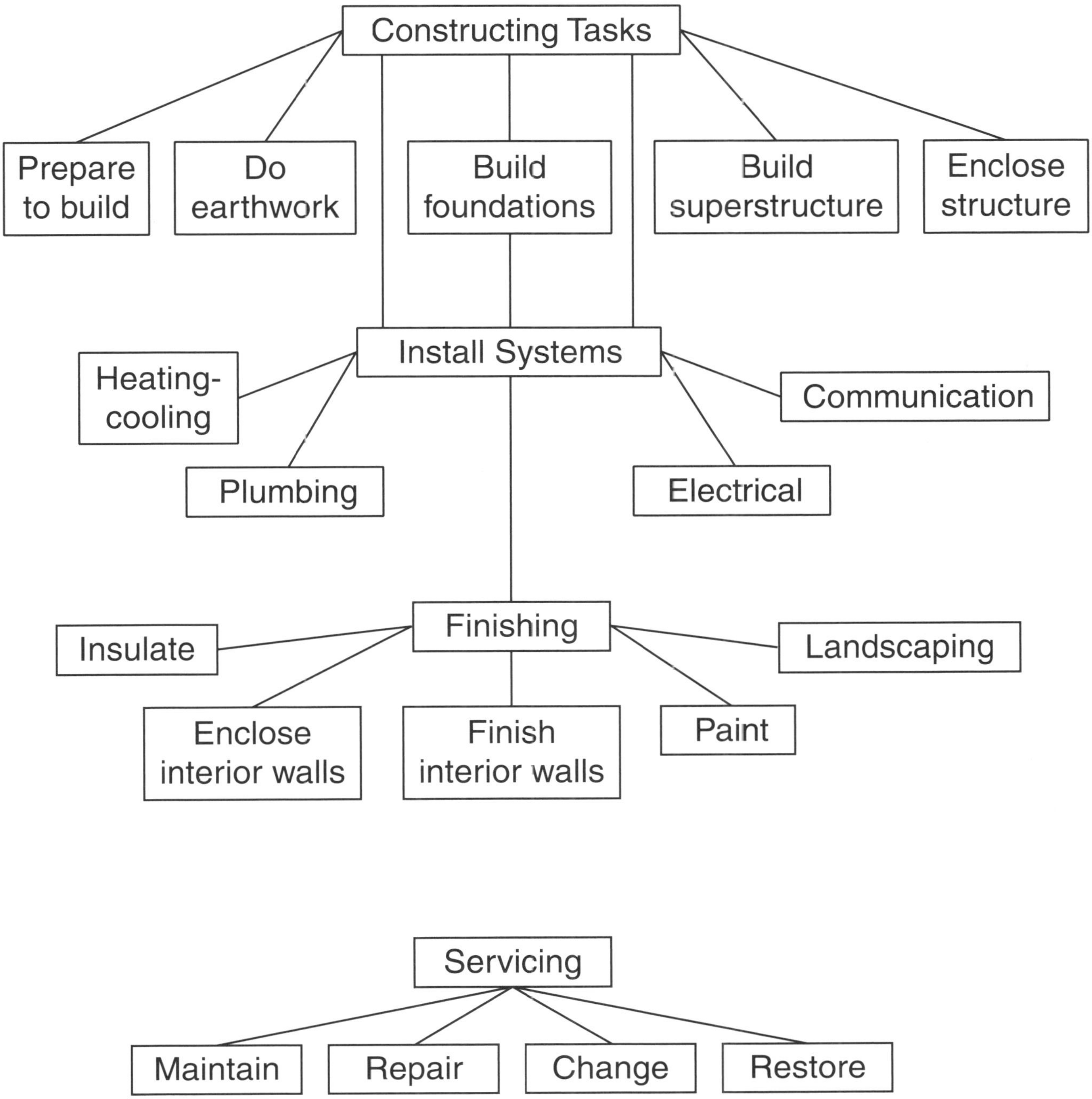

Reproducible Master 20-4

Energy Conversion
Steam Power Plant

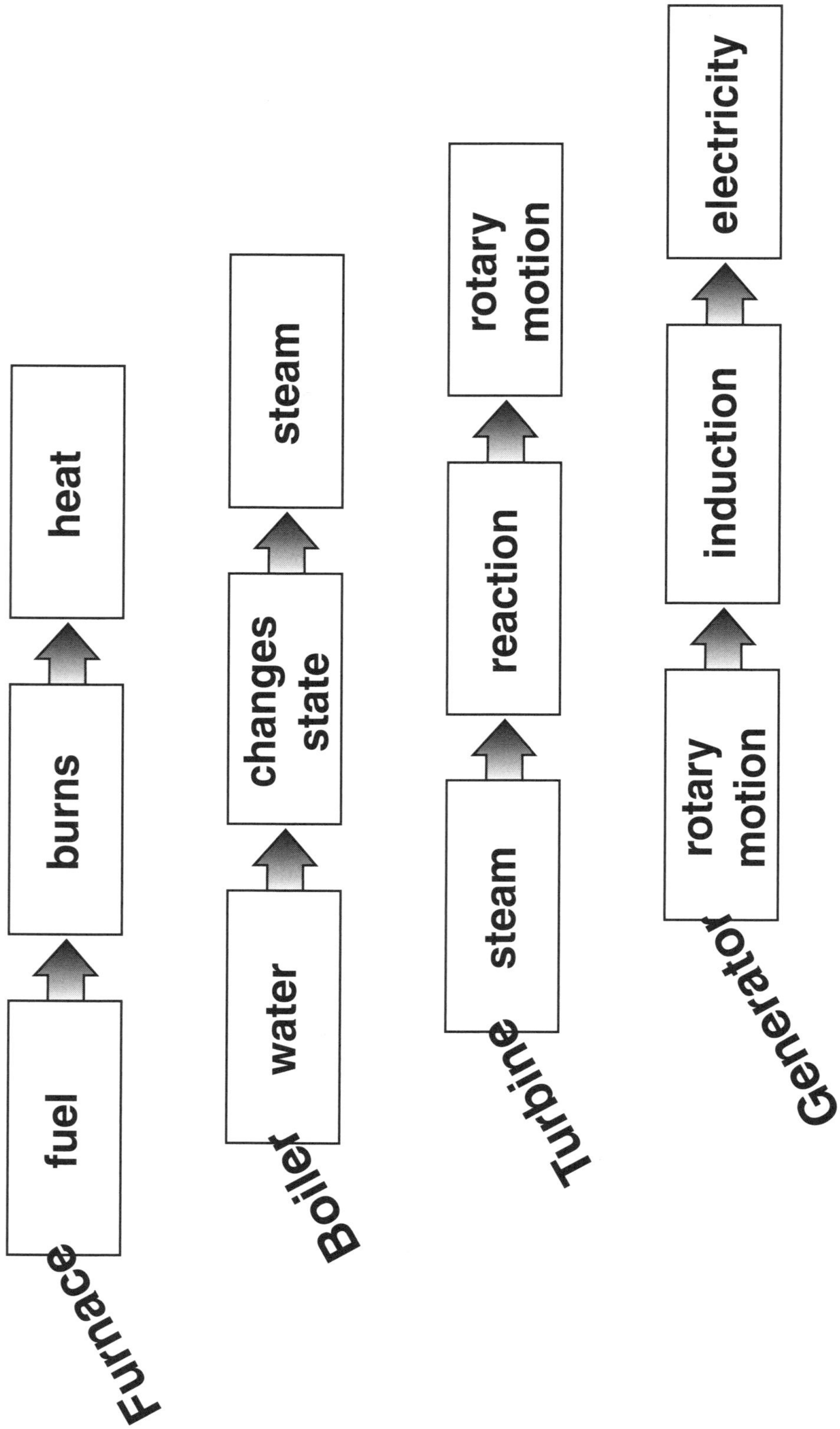

Reproducible Master 21-1

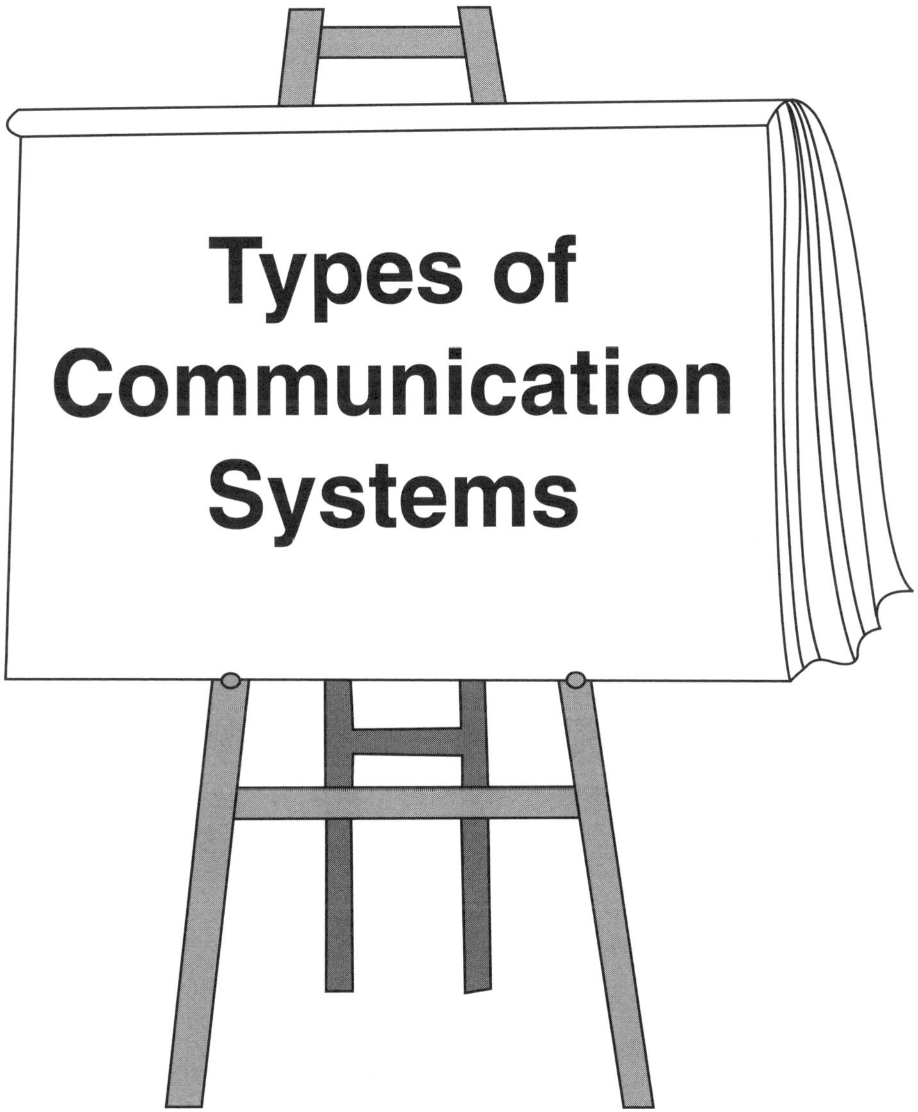

- Printed graphic communication
- Tele-communication (electronic)
- Photographic communication
- Technical graphic communication

Reproducible Master 22-1

The Communication Process

Encode - Transmit - Receive - Decode

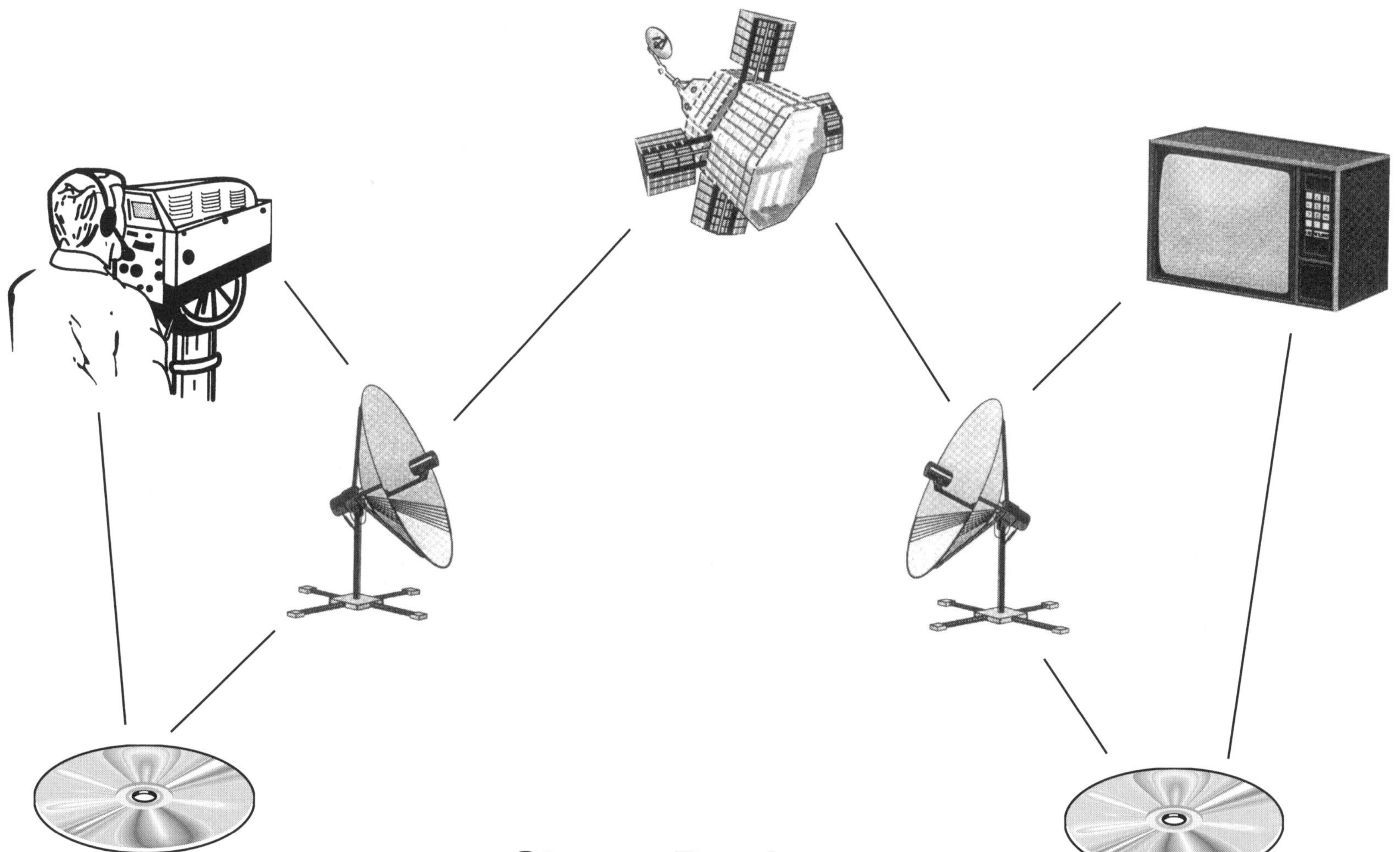

Store - Retrieve

Creating Media Messages

1. Designing the message.
2. Preparing to produce the message.
3. Producing the message.
4. Delivering the message.

Reproducible Master 22-3

Steps Used to Produce a Message for Print Communication

1. Designing the message.
   - ❑ Determine who is the audience (audience assessment).
   - ❑ Find out what gets the attention of the audience.
   - ❑ What size and shape will the message carrier be? Will the carrier be a flyer, billboard, or newspaper (tabloid or full-size)?
   - ❑ Collect information for the message.
   - ❑ Write the copy.
   - ❑ Select illustrations (drawings and photos).
2. Preparing for production.
   - ❑ Make rough sketches.
   - ❑ Make refined sketches.
   - ❑ Make comprehensive sketches.
3. Producing the message.
   - ❑ Schedule printing.
   - ❑ Set type. Size illustrations and photos.
   - ❑ Prepare a layout.
   - ❑ Prepare the image carrier.
   - ❑ Print the message.
4. Delivering the message.
   - ❑ Distribute the magazine, newspaper, or flyer.
   - ❑ This might involve mailing the message or contacting private distributors (stores offering books and magazines for sale).

Reproducible Master 22-4

Designing a Message

Printed Graphic Media

1. Assess the audience.
2. Establish the format.
3. Gather information.
4. Write and edit copy.
5. Design graphics.

Telecommunication Media

1. Assess the audience.
2. Establish the format.
3. Gather information.
4. Write and edit scripts.
5. Design graphics.

Reproducible Master 22-5

Reproducible Master 22-6

Preparing to Produce a Message

Printed Graphic Media

1. Design the layout.
2. Generate the copy.
3. Generate graphics.
4. Prepare the electronic layout.

Telecommunication Media

1. Schedule production.
2. Cast.
3. Stage.
4. Rehearse.

Producing a Message

Printed Graphic Media

1. Produce the image carrier.
2. Transfer the image.
3. Finish the product.

Telecommunication Media

1. Record images.
2. Edit the recording.
3. Produce the master.

Delivering a Message

Printed Graphic Media

1. Pack the product.
2. Ship to customers.
3. Read or store the message.

Telecommunication Media

1. Broadcast the message.
2. Receive the message.
3. View, listen to, or store the message.

Reproducible Master 22-8

Parts of a Computer

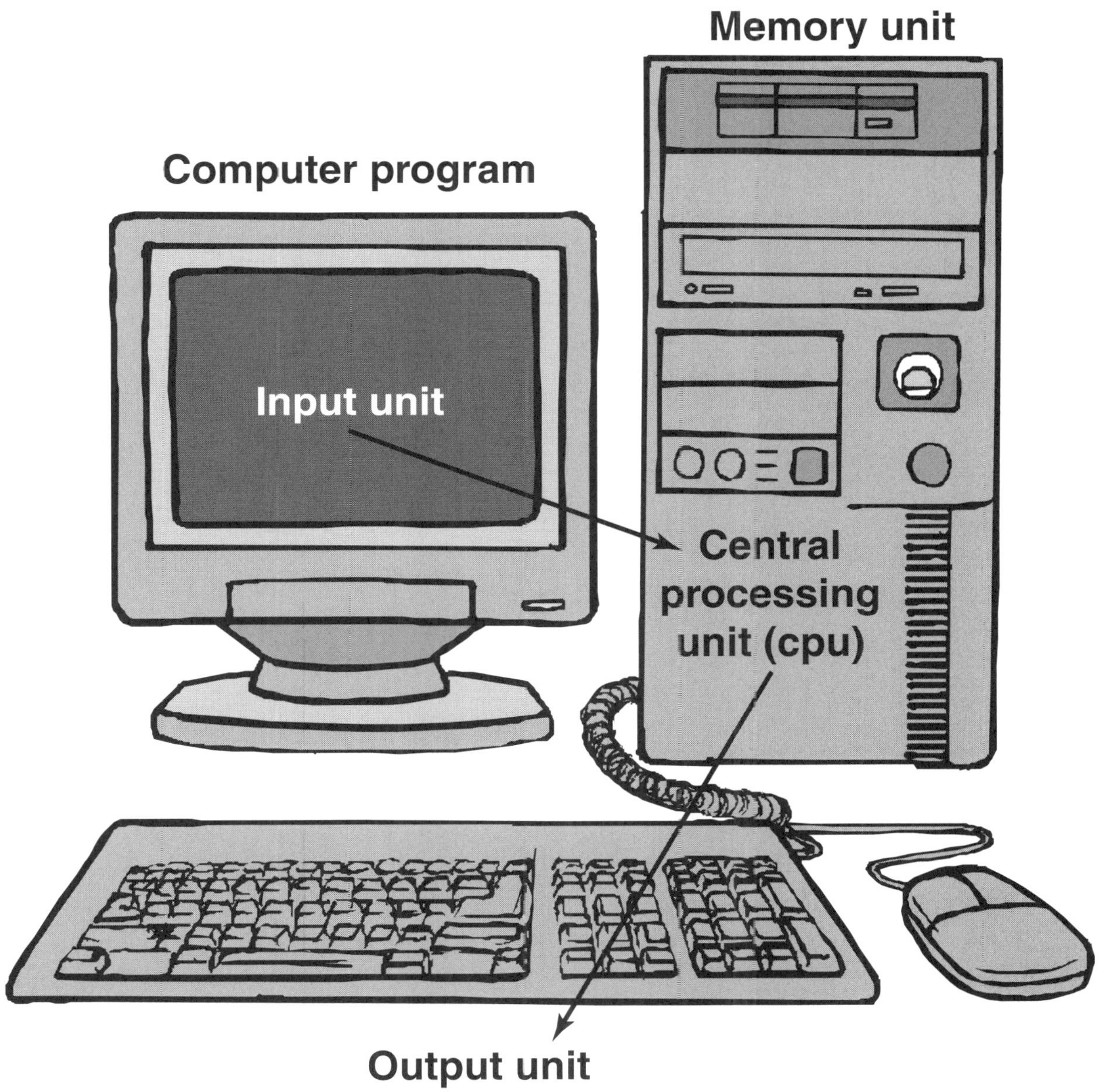

Reproducible Master 22-9

Communication Guidelines

The following sections provide guidelines in a variety of communication media. Following these guidelines will help to ensure that your communication messages meet the receiver's expectations.

Word Processing

Probably the most common form of communication you encounter in school and in the field of technology is written communication. Most written communication is produced using word processing. Word processors are computer-software programs that allow you to create neat and easy-to-read documents. It is up to you to make sure the content is organized and accurate.

When typing a document, there are several guidelines to follow. First, always use correct grammar and punctuation. If you are unsure about what is correct, use a dictionary or other resource to find out. Next, avoid using extra returns to skip lines between paragraphs. Similarly, use only one space after each sentence. To modify the visual effects of your document, you can change the font type or size, justify the text, add tabs, or use a different text style.

Word processors help you create electronic files. These files should be carefully saved for later use. Normally, the files can be saved on removable media (flash drive, CD, or digital videodisc) or the computer's hard drive. It is wise to save your work often and to even save a backup copy so you are sure not to lose it.

Communicating with written documents created on a word processor is most effective if your message is organized, accurate, and clear. For this reason, it is important to proofread your work before finalizing it. Often, several drafts need to be written before the document is acceptable for printing. Once the document is finished, proofread, edited, and saved, you can usually print it out on a printer connected to your computer.

Video Communication

Similar to any other form of communication, video requires planning and organization. When outlining your message, remember that videos are largely visual. The pictures you film are often more important than the words you write. To help organize your visual images, it is useful to create a storyboard. A storyboard is a series of pictures that tell a story, very similar to a comic strip. Once your pictures are created, you can try rearranging them for different effects and decide on a final sequence to use in your video. In addition to visual images, your video will probably also require a script. The spoken words follow the sequence of pictures and help tell your message.

After the initial planning is complete, it is time to begin planning for the production of your video. Scenes are not always recorded in the sequence in which they will appear in the final product. The order in which you film the scenes might depend on the availability of set locations, groups of actors, and special props. To help ensure your final product is interesting to watch, you should plan to record your subject from various angles and use a wide range of shots. Also try to shoot cutaway shots. Cutaway shots are filler shots that can be used for transitions.

Video communication does require specialized equipment. There are many types of video cameras. You want to use one that best suits your needs. Extra batteries and handheld microphones might also be needed.

(Continued)

Once you have recorded your scenes, there are numerous effects you can use to enhance your video message. Video-production software is useful for applying some of these effects to your recording. Both visual effects and sound effects can be added to your video. Voice-over narration can be added during postproduction. Your cutaway shots can also be inserted to create a smooth, flowing final product.

PowerPoint Communication

Oral presentations can be greatly enhanced by incorporating a PowerPoint presentation. Content and design are both important when dealing with the PowerPoint presentation graphics program, so it is essential that you research your topic well and plan out a general design for your PowerPoint slides. Your slide design should be kept simple. The font size should be large enough to read from the back of the room where you will be giving your presentation. Each slide should include a title and a single piece of information or a short list of bulleted points. The information you present in your oral presentation should only be outlined on the PowerPoint slides, not written word for word.

Web-Based Communication

Web-based communication is becoming increasingly popular and important in technological industries. Developing a Web site takes a good deal of planning and time. Sketch a plan for each Web page. Make sure you include all the necessary links between pages.

The most important characteristics of a good Web site are ease of navigation and functionality. It is important that your site has a nice visual appearance. Easy-to-read type and easy-to-find links, however, are more important. Use common fonts. Limit the types and sizes of fonts as much as possible. You can add images where they might be appropriate. Using too many images, however, will cause the page to load slowly and might detract from your intended message. To create a successful Web site, try to keep things simple and clear.

Desktop Publishing Communication

Communicating with desktop publishing software is a strictly visual activity, so the principles of design are especially important. When designing a message for printed communication, the various elements of the message should be balanced and proportionate. Contrast, unity, and rhythm can be used to achieve the desired effect.

Font types and sizes should be kept to a minimum to avoid confusing your audience. Choose your illustrations carefully. Do not forget the effects that the use of white space can have on your final product. Try arranging the elements of your design in several ways before deciding which way is most effective for your purpose. Make sure to draw particular attention to any titles or other essential information. The visual design of your message determines how the reader's eyes move when viewing your finished product.

Secondary Manufacturing Processes

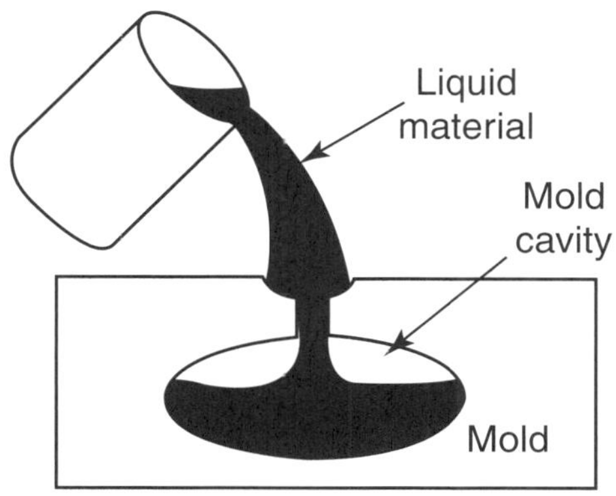

Casting and Molding

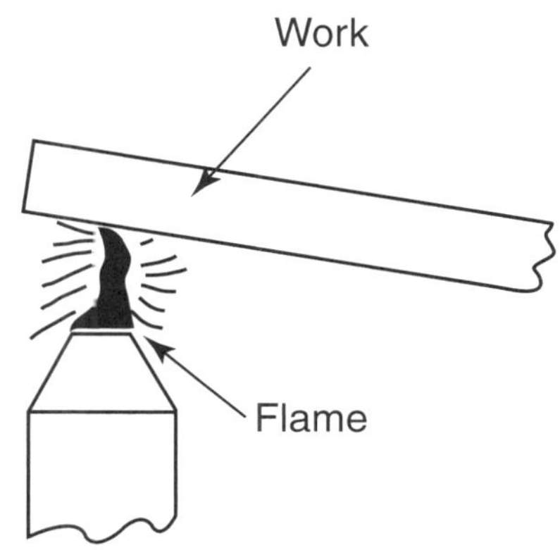

Conditioning

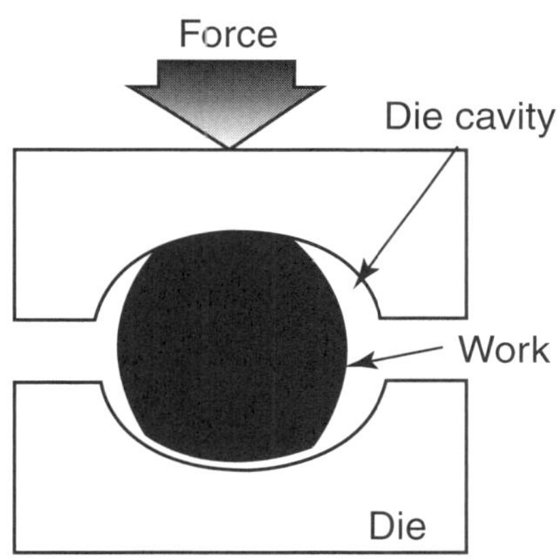

Forming

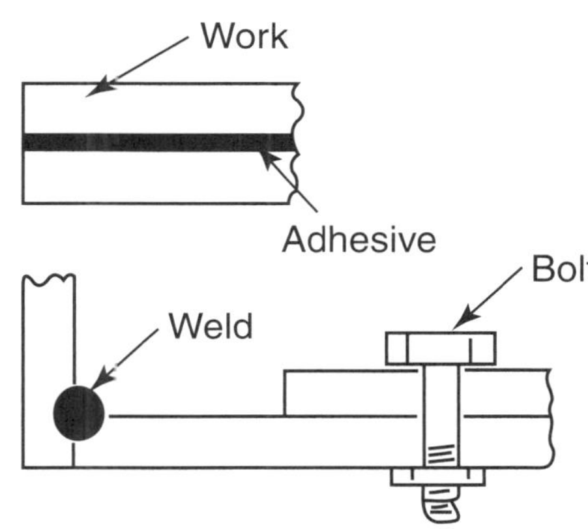

Assembling

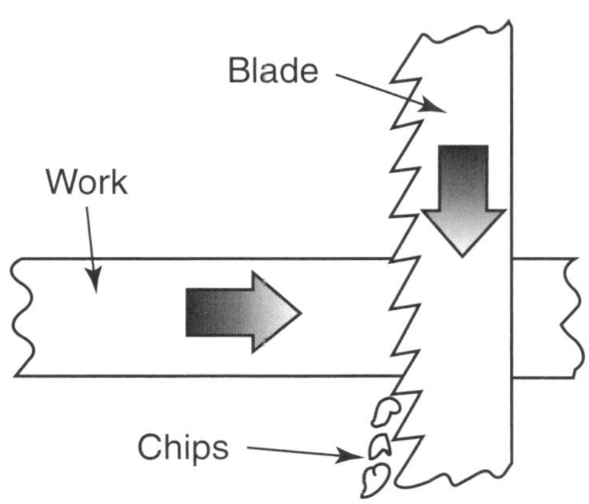

Separating

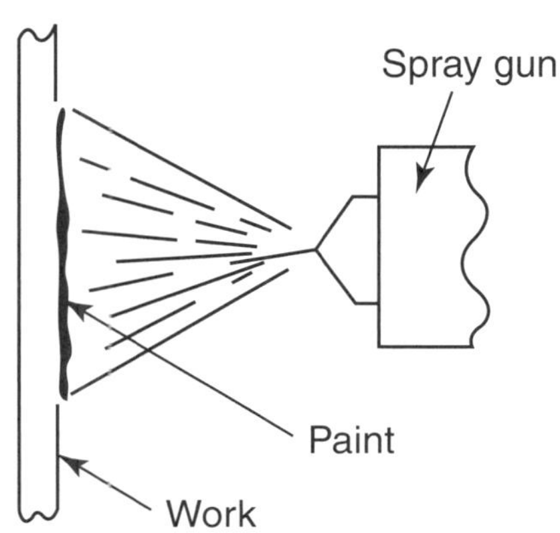

Finishing

Reproducible Master 23-1

Technology and Medicine

People use technology in health and medicine to do the following:

- ❑ Help prevent illness and injury.
- ❑ Diagnose diseases and injuries.
- ❑ Treat illnesses and injuries.

Wellness

Wellness—State of personal well-being

Wellness programs have people do things that help keep their bodies healthy.

Reproducible Master 24-2

Vehicle Systems
Structure

Provides the frame and covering for the major areas of the vehicle.

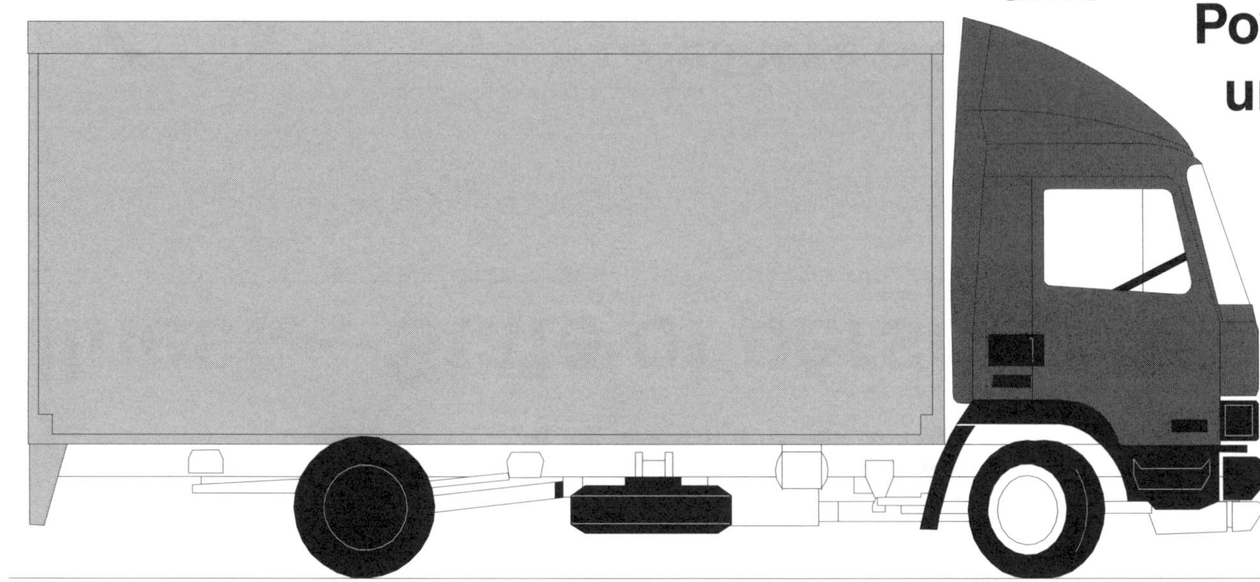

Vehicle Systems
Propulsion

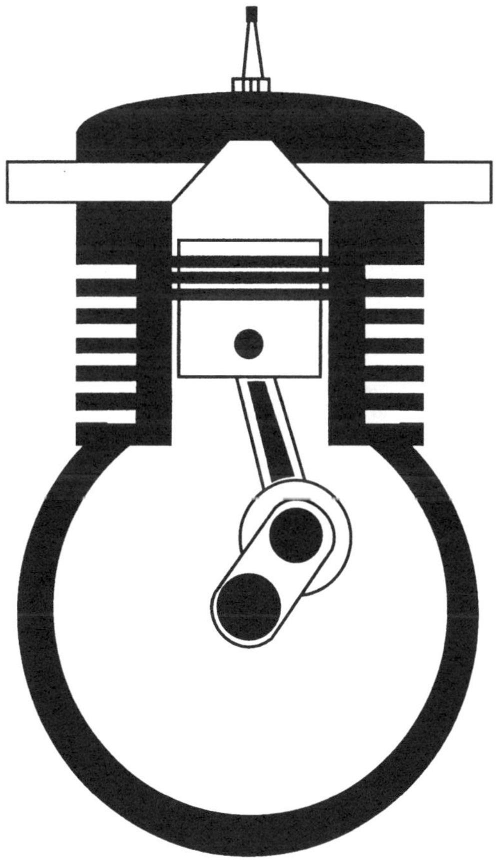

Converts energy into power to move the vehicle.

- ❑ Power generation
- ❑ Power transmission

Reproducible Master 25-2

Vehicle Systems
Suspension

Supports the weight of the vehicle and the cargo.

- ❑ Mechanical (wheel and axles).
- ❑ Fluids (air, oil, and water).

Reproducible Master 25-3

Vehicle Systems
Guidance

Provides information needed to control the vehicle's speed and direction.

- ❑ Vehicle location
- ❑ Vehicle speed
- ❑ Operating conditions
- ❑ Rules and laws

Name ______________________________ Date ___________

Period _____________________________ Score ___________

Vehicles and Their Systems

On the chart below, list the various devices that make up the five systems. (Look at the example. Fill in the boxes.)

Vehicle	Propulsion	Transmission	Guidance	Control	Measurement
Bicycle	Rider	Pedals Crank Sprockets Chain	Signs Traffic light	Handlebars Hand brakes Switches	Speedometer Odometer
Auto					
Airplane					
Mousetrap					
Race car					
Golf cart					
Blimp					

Reproducible Master 25-5

Section 5
Technology and Society

Focus: Exploration of the impacts of technology on resources, people, and the environment and how change will affect the future of technology.

- ❑ Chapter 26: Technological Impacts
- ❑ Chapter 27: Technology and the Future

Teaching Materials

Text

Chapter 26, pages 628–655
- Chapter Activities, page 650
- Test Your Knowledge, page 650
- Chapter 26 Activity A— Technological Impacts, pages 652–653

Chapter 27, pages 656–669
- Chapter Activities, page 668
- Test Your Knowledge, page 669

Tech Lab Workbook

Chapter 26
- Student Activity 26-1—Impacts of Technology, page 145
- Chapter 26 Study Questions, page 147

Chapter 27
- Student Activity 27-1—Future Telling, page 149
- Student Activity 27-2—Products of the Future, pages 151–152
- Chapter 27 Study Questions, page 153

Teacher's Resources

Chapter 26
- Reproducible Master 26-1—Global Warming, page 191
- Reproducible Master 26-2—Technology Assessment, page 192
- Lesson Slide 26-1—Impacts of Technology (CD only)
- Lesson Slide 26-2—Energy Resources (CD only)

Chapter 27
- Reproducible Master 27-1—Modeling a Major Event, page 193

Introduction

This section is placed at the end of the course, not as an afterthought, but as a planned conclusion to the study of technology. Knowledge of the impacts and future of technology is vital to the understanding of how society uses and controls technology. Technological resources, designed products, and applications are only useful if their impacts on society are understood. To be technologically literate, one must understand how technology functions in society. It is equally important to understand that we, as a society, control the directions and outcomes of technology. By understanding technology's impacts and recognizing where technology is headed, we choose what technological changes to make. Only a technologically literate society has this power.

Objectives of the Section

At the end of Section 5, each student should be able to do the following:

- ❑ Give examples of technological change.
- ❑ Understand the concept of a negative impact.
- ❑ Explain impacts on people, resources, and the environment.
- ❑ Summarize ways in which the future can be predicted.
- ❑ Give examples of future technology.

Suggested Schedule and Length of the Section (5–10 Days)

Section 5 introduces the students to the topic of technology and society. Have them complete the following assignments:

1. Read Chapters 26 and 27 in their textbooks and answer associated study questions.
2. Participate in several discussions and presentations.
3. Complete the scheduled activities and follow-up sheets.

Schedule for Section 5— 9-Week, 12-Week, 18-Week, and 36-Week* Classes	
Day 1	Discussion: "Technological impacts." Assign Chapter 26: "Technological Impacts" to be read outside of class. Introduce first activity, "Technological Impacts."
Day 2	Complete and summarize activity. Discussion: "Forecasting the future." Assign Chapter 27: "Technology and the Future" to be read outside of class.
Day 3	Discussion: "Future technologies." Introduce second activity, "Products of the Future."
Day 4	Work on activity.
Day 5	Complete and summarize activity. Summarize course.

* For a 36-week class, use an additional activity and spread the discussions over the two weeks scheduled for this session.

Preparing to Teach the Section

Before introducing this section, complete the following tasks:

1. Study this section of the Teacher's Resource.
2. Read Chapters 26–27 in the textbook to review the content.
3. Prepare three 10- to15-minute discussions and presentations on the following topics:
   A. Technological impacts. (Textbook, Chapter 26)
   B. Forecasting the future. (Textbook, pages 657–663)
   C. Future technologies. (Textbook, pages 663–667)
4. Select appropriate laboratory activities from the following options:
   A. Technological Impacts (Chapter 26), Textbook, pages 652–653.
   B. Impacts of Technology (Chapter 26), Tech Lab Workbook, page 145.
   C. Future Telling (Chapter 27), Tech Lab Workbook, page 149.
   D. Products of the Future (Chapter 27), Tech Lab Workbook, page 151.
5. Prepare materials and supplies for the selected activities and presentations.
6. Work through the activities and student study questions presented in the textbook and Tech Lab Workbook.

Presenting the Section

This section should be taught by using the three short discussions to reinforce the concepts from the textbook. The activities should be used to support the content from the discussions and readings. The students' understanding of the concepts can be evaluated using the Test Your Knowledge questions in the textbook and study questions in the Tech Lab Workbook. At the end of the section, it is important that the section is summarized to highlight the importance of understanding the role of technology in society. The last day of this section should also be used to review and summarize the entire course. Prepare a presentation emphasizing the key points of each of the five sections the students have studied.

Adjustments for Students with Special Needs and Gifted Students

The activities in this section are open-ended. They will be easier for the gifted students to handle than the prescriptive activities found in some of the other sections. Give special attention to students who might require assistance in generating ideas and creative solutions. All these activities lend themselves to either individual work or group work. Consider using one or more of these activities to combine students of varied abilities. For example, if the "Future Telling" activity is done in a group, it might be beneficial to pair a student with good writing skills and a very imaginative student with low writing ability.

Evaluation Procedures

This section stresses the understanding that technology impacts society and that technology's future can be forecasted. The focus should be on the general understanding of the different types of impacts and forecasts. There are no right answers about the impacts and future of technology. Each student might have different opinions and feelings about how technology has affected and will continue to affect society. Students should be made to feel that their views are not the only views of technology and that even opposing views can be correct. This makes evaluation of this section subjective. Therefore, students should be evaluated on the following factors:

1. Contributions to group activities. This can be done with daily logs (diaries) the students maintain, teacher observations, and student feedback.
2. Laboratory sheets, sketches, and designs completed during the activities.
3. Participation in discussions conducted during the activities.
4. Scores on quizzes and chapter questions.

Implementing the Laboratory Activities

Three different activities have been developed for this section. They can be found at the end of Section 5 in the textbook and in the Tech Lab Workbook. These activities are suggested activities introducing the students to concepts relating to technology and society. Any activity on the topic of technological impacts or the future of technology can be used in place of any of these.

Procedures for Directing Each Suggested Activity

Chapter 26: Technological Impacts, Textbook, pages 652–653

This is an activity allowing students to explore how the technology they use affects society. The activity is a one- to two-day activity that works best if the students work individually or in pairs. They will select everyday objects from around the classroom and explore what happens to the objects when they are discarded. The students will list the ways the objects impact society. To conclude the activity, they will select one of their objects and create a drawing of changes that can be made to decrease the negative impacts.

Chapter 26: Impacts of Technology, Tech Lab Workbook, page 145

In this activity, the students will select a technology and determine a number of impacts related to its use. Students will need access to research materials (library or Internet access). It might be helpful to provide an example of this activity you have generated for the students. To provide options for this activity, you can have students select a previously developed technology and research the actual impacts that have occurred, or the students can select an emerging technology and hypothesize about the potential impacts.

Chapter 27: Future Telling, Tech Lab Workbook, page 149

"Future Telling" is an alternative activity to be used if time permits. This activity is designed to be creative and open-ended. The students look into the future and create a scenario of a future event. They should be encouraged to be imaginative and creative. This activity is an individual project. Groups can be used, however, to help the students generate ideas.

Chapter 27: Products of the Future, Tech Lab Workbook, pages 151–152

This is a group activity in which the students forecast the technology that will be available in a certain area of technology in the future. The students will be assigned one context of technology. They will examine the products and systems currently available in that context. The students will then create drawings of a technology that might exist in the future.

1. Divide the class into seven groups.
2. Give each group one of the seven contexts of technology: agricultural and biorelated technology, construction technology, energy-conversion technology, information and communication technology, manufacturing technology, medical technology, or transportation technology.
3. Have the students generate a list of current technologies in their areas. They should record them on the activity sheet on page 151 of the Tech Lab Workbook. The students should use the textbook, library, and Internet for ideas.
4. They should then brainstorm to come up with future technologies in their areas. Space is provided for students to write down their ideas during brainstorming.
5. Have each group choose one of the ideas and develop it further. The students are to complete the drawing and explanation portions of the worksheet in the Tech Lab Workbook.
6. If time permits, in a 36-week course, supply the students with the materials and supplies to build models of their technologies.
7. To summarize the activity, have the students present their technologies to the rest of the class.

Answer Key for Section 5

Chapter 26—Technological Impacts

Test Your Knowledge, Textbook, page 650

1. Answers will vary. Some acceptable answers are to improve the world around us, to do work more easily, to travel more easily, to have more time to enjoy ourselves, and to help others.
2. These materials are developed to meet a specific need.
3. Answers will vary. Some acceptable answers are a reduction in the amount of labor needed to complete a job and the production of higher-quality, more consistent products.
4. The feeling that we are closely connected to everyone on the globe.
5. We can now travel much faster and much farther, in more comfort.
6. Technology has given us new materials and building techniques.
7. productive
8. Answers will vary. Some acceptable answers are reduce the use of nonrenewable resources, develop and use renewable resources, and promote the use of inexhaustible resources.
9. Answers will vary. Some acceptable answers are air pollution, water pollution, noise pollution, and soil erosion.
10. technological unemployment

Study Questions, Tech Lab Workbook, page 147

1. change
2. False
3. C. Materials.
4. D. Tools.
5. B. Communication.

6. A. Construction.
7. Answers will vary. Some acceptable answers are that technology can create scarcity, it can repair damage natural disasters cause, and it can replace people in the workforce.

Chapter 27—Technology and the Future

Test Your Knowledge, Textbook, page 669

1. Answers will vary. Some acceptable answers are that there would be no Internet, television, automobiles, microwave ovens, or coffeepots that strained the grounds.
2. futurist
3. Answers will vary. Some acceptable answers are economists, writers, and meteorologists.
4. mid-range
5. Trending, surveying, creating scenarios, and modeling.
6. D. All of the above.
7. False
8. The event is explained. A possible solution to the problem is developed. The expected outcome of the solution is determined.
9. Answers will vary. Some acceptable answers are hydroponics, genetically altered crops, and more automated farming processes.
10. False

Study Questions, Tech Lab Workbook, page 153

1. futurist
2. forecasts
3. 50
4. B. Trending.
5. A. Surveying.
6. C. Developing a scenario.
7. medical
8. manufacturing

Captions for Reproducible Masters

- ❑ **Reproducible Master 26-1.** Global Warming—How global warming works.
- ❑ **Reproducible Master 26-2.** Technology Assessment—A chart for students to fill out when studying the positive and negative impacts technology can have on our world.
- ❑ **Reproducible Master 27-1.** Modeling a Major Event—A worksheet for students to complete when learning about modeling events.

Captions for Lesson Slides (CD Only)

- ❑ Lesson Slide 26-1. Impacts of Technology—Technological devices impact people and the environment.
- ❑ Lesson Slide 26-2. Energy Resources—Our limited supply of exhaustible resources will require us to rely more on renewable and inexhaustible resources in the future.

Global Warming

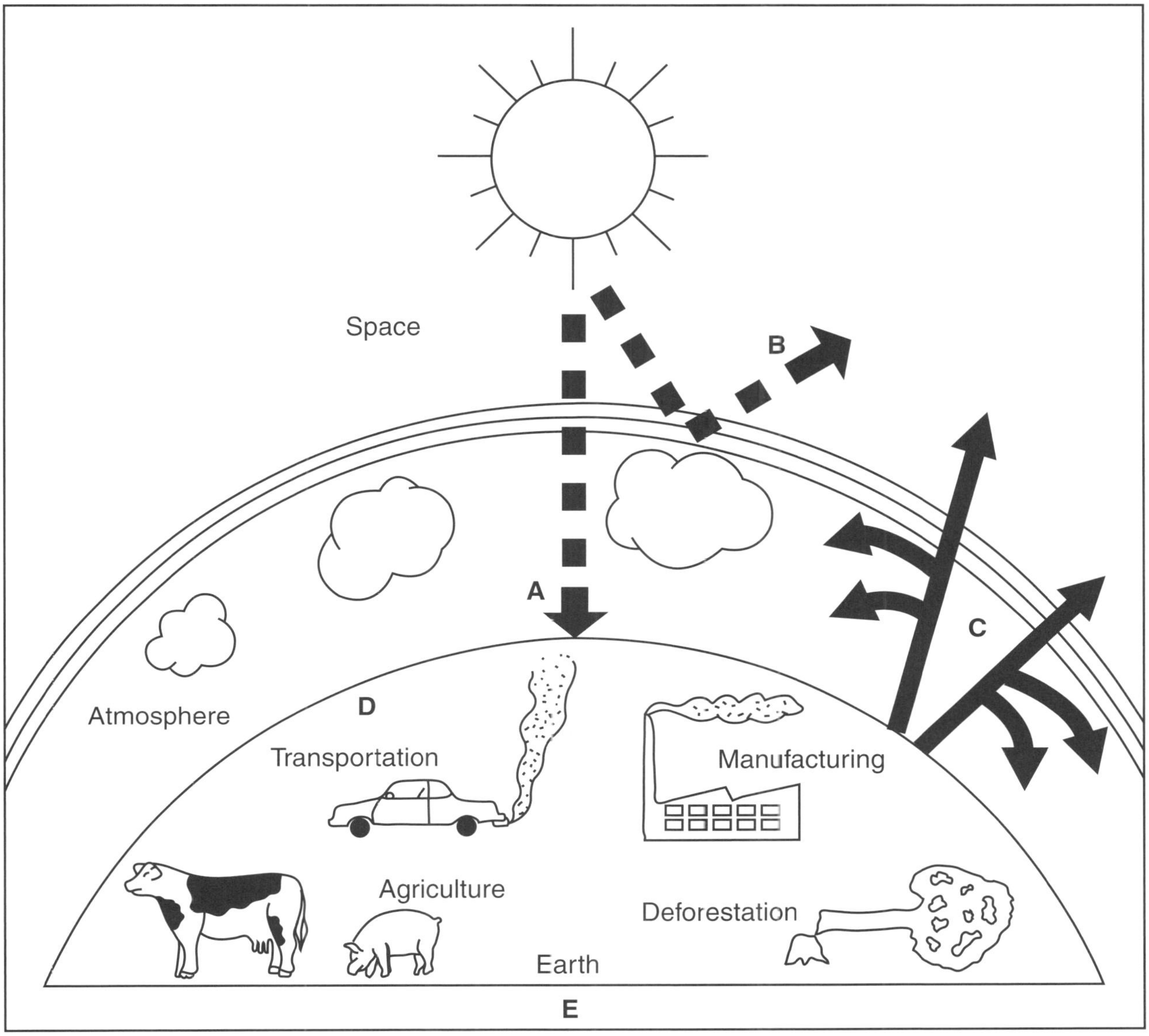

A. About 130 watts per square foot of solar energy reach the Earth's atmosphere at any one moment.
B. Around 30% of this energy reflects back into space.
C. An ozone layer high above Earth shields Earth's inhabitants from harmful ultraviolet solar radiation.
D. The use of fossil fuels for energy and the destruction of Earth's forests are major reasons for increased carbon dioxide in the atmosphere. Other pollutants are methane from swamps, landfills, and cud-chewing animals. Chlorofluorocarbons (synthetic gases used in refrigerants, spray cans, and insulation) are reducing the protective ozone layer.
E. All these pollutants together form a cover in the atmosphere. This cover traps heat energy, thus warming Earth.

Reproducible Master 26-1

Technology Assessment

As you have learned, technology, while often making our lives more comfortable and enjoyable, can have undesired effects as well. In the chart below, put the name of a product or structure you use frequently. List the desired effects of the product or structure in the first column. In the second column, list the undesired effects.

Name of Product or Structure: ________________	
Desired Effects	**Undesired Effects**

Reproducible Master 26-2

Modeling a Major Event

In the spaces below, prepare a model of an event (such as a dance or fundraiser) in your school. Decide what the major event will be. List all the minor events that must take place to make the major event a success.

Major Event

Name of event: ____________________

When it will take place: ____________________

Minor Events

Number 1

Name of event: ____________________

When it will take place: ____________________

Number 2

Name of event: ____________________

When it will take place: ____________________

Number 3

Name of event: ____________________

When it will take place: ____________________

Number 4

Name of event: ____________________

When it will take place: ____________________

Number 5

Name of event: ____________________

When it will take place: ____________________

Number 6

Name of event: ____________________

When it will take place: ____________________

(Continued)

Reproducible Master 27-1

Number 7

Name of event: ____

When it will take place: ____

Number 8

Name of event: ____

When it will take place: ____

Number 9

Name of event: ____

When it will take place: ____

Number 10

Name of event: ____

When it will take place: ____

Reproducible Master 27-1